Weed Management for Organic Farmers, Growers and Smallholders

Weed Management for Organic Farmers, Growers and Smallholders

A COMPLETE GUIDE

Gareth Davies, Becky Turner and Bill Bond

In association with the Henry Doubleday Research Association

THE CROWOOD PRESS

First published in 2008 by
The Crowood Press Ltd
Ramsbury, Marlborough
Wiltshire SN8 2HR

www.crowood.com

British Library Cataloguing-in-Publication Data
A catalogue record for this book is available from the British Library.

ISBN 978 1 86126 970 6

Typeset by Exeter Premedia Services Private Ltd., Chennai, India

Printed and bound in India by Replika Press

Contents

	Acronyms, Abbreviations and Plant Names	6
	Preface and Acknowledgements	7
Chapter 1	Principles of Weed Management on Organic Farms	9
Chapter 2	Preventative and Cultural Weed Management	24
Chapter 3	Direct Weed Control	61
Chapter 4	Economics of Weed Management (*with Ulrich Schmutz*)	95
Chapter 5	The Weeds	113
Chapter 6	Weed Management Strategies in Systems and Crops	196
Chapter 7	Solving Weed Management Problems	246
	Appendix: Useful Information for Weed Management	259
	Index	268

Acronyms, Abbreviations and Plant Names

PLANT NAMES

The first time a weed or crop is mentioned by name we give the common UK name followed by the standard botanical (or *latin*) name to provide an absolute reference point. This is because common names vary widely between regions. Thereafter the same common name is used. Full Latin names are also provided for all weeds specifically described in Chapter 5.

ABBREVIATIONS

ac	acre
cm	centimetre(s)
ft	foot
FYM	farmyard manure
ha	hectare
in	inch(es)
km/h	kilometres per hour
LAI	leaf area index
m	metre
MJ	mega joules
mph	miles per hour
t	metric tons

ACRONYMS

EU	European Union
Defra	(UK) Department of Environment, Food and Rural Affairs
IFOAM	International Federation of Organic Agriculture Movements
NIAB	(formerly) National Institute of Agricultural Botany
UK	United Kingdom

Preface and Acknowledgements

On any farm walk, discussion will inevitably turn at some point to weeds. In all organic crops and rotations, weeds are always a background presence, and farmers and growers all have stories to tell about them, how they manage them and, ultimately, how they come to accommodate them. This reflects the truth that weeds will always be present, and it will always be necessary to manage them in order to avoid or mitigate their deleterious effects. But, this is also a reflection of another underlying truism, which is that weeds are a consequence of human intervention in, and human perceptions of, the ecological setting in which agriculture is practiced. From these (sometimes very different) perspectives, weeds do not only have negative consequences, but can also have neutral or even beneficial effects, which can reach beyond the field boundary to take in landscape or social values. Organic agricultural principles specifically encourage farmers and growers to take these latter effects into account in their farming practices; without, of course, losing sight of the damage that they can cause.

Organic weed management, therefore, becomes the property of a whole rotation or farm system, and not just a matter of a limited interaction between a crop and a weed in a specific setting. This has important consequences for weed management on organic farms because 'problems' arising in such a context are unlikely to 'solved' by a technological fix at some specific point, applied according to schedule, and then forgotten. Rather they require a consistent long-term approach, adapted to the specific environmental and socio-economic setting of any particular farm, which aims to manage the weeds, drawing out any positive effects they might have and suppressing any negative ones. In fact, in any specific situation, there will a wide range of factors relevant to taking decisions about weed management, from the immediate to the long term, and this has led to a wide diversity of weed control practices and techniques on organic farms.

In this book we have attempted to bring together all the most important factors that impinge on weed management decisions on organic smallholdings and farms, and to build up a picture of how these practices and techniques can be integrated to build such a consistent approach to managing and controlling weeds, while adhering to organic principles. This has necessitated a broad approach. Whilst weed management techniques in organic farming can draw on the most modern scientific theory and technology, such as plant competition ecology and computer-assisted hoes, they also rely on tried and trusted technologies and practices that have been developed by farmers and growers over generations, a process that is on-going into the present time, sometimes with the aid of advisors and researchers, sometimes not. It is in this spirit that we offer this book, as a snapshot of ideas and practice for organic weed management, in the hope that it will provide a valuable source of information for those wishing to control weeds on their farms or holdings, but also serve as a springboard for those wishing to develop and improve their organic weed management practices. May we be able to rewrite the book completely in fifteen years time!

Acknowledgements

The authors wish to thank the numerous people involved in helping to bring this book to publication, especially Phil Sumption, for his advice on weed management in early drafts, and for bringing his editing skills to bear on the final draft. In bringing a book like this to publication we have drawn upon our experience in organic weed research over many years and have relied upon the knowledge, experience and advice of many people in the 'organic movement' and beyond, without whom we could not have filled the pages. Any mistakes are ours not theirs. Within this group a special mention must go to the numerous farmers and growers whose experience in organic weed management forms the backbone to this book, and we especially would like to say thank you to them for having the patience to explain why and how they carry out weed control on their farms, and for sharing their experiences both with us, and with their colleagues. Without these stories, our understanding of weed management would be meagre. A final acknowledgement must go to Defra whose funding for organic research has helped to underpin much of our work at HDRA.

Chapter 1

Principles of Weed Management on Organic Farms

Weeds are usually defined as any plants growing where they are not wanted. In an agricultural setting, weeds are traditionally taken to mean any non-crop plants in a field, but could also include plants growing in field margins and in other locations where they are likely to spread or otherwise cause a nuisance. They are normally perceived to be doing some 'damage' or 'harm'; that is, they are having negative effects like reducing yields, lowering crop quality, causing difficulties with crop management operations or even poisoning livestock. Weeds are, to a greater or lesser extent, ecologically adapted to man made and agricultural habitats, and for this reason are able to persist, and even multiply, if steps are not taken to manage them. It is the task of this book to provide an overview of weed management for organic farmers, growers and smallholders. It will also be suitable for those wishing to reduce their use of synthetic chemical herbicides.

In order to manage weeds successfully, it is necessary to understand both their life-cycles and their ecological characteristics, although this in itself is not sufficient. It is also necessary to evaluate the circumstances in which they are to be controlled, in order to identify the points at which they are most vulnerable in the whole farm system. In this chapter we develop a framework for approaching weed management in organic farming systems by exploring what weeds are, why farmers regard weeds as 'problems' and describing the context in which they need to be managed. In subsequent chapters we expand upon both the implications of these ideas and the methods available to organic farmers and growers to manage weeds, by describing in more detail the methods useful for preventing weed populations building up and the measures available to more directly control them. It is important to understand the likely costs of controlling weeds in order to offset this against any potential benefits of leaving them, and we present ideas on how to assess these costs and benefits. We also provide a summary digest of the weeds that occur, and present ideas as to how to plan for weed management and how to incorporate weed management into rotational and cropping strategies.

WEEDS ON ORGANIC FARMS

Farmers and growers try and control weeds because they have a direct negative impact on their farm businesses; for example, by reducing yields or increasing costs. Weeds also have less direct and overlooked negative effects, not least on a farmer's feeling of well-being or perception of how 'well' they are farming compared to their neighbours. For these and other reasons, all farmers manage 'unwanted' vegetation, or weeds, on their farms. Notwithstanding this, it is also important to realize that the term 'weed', or the definition of a particular plant as a weed, is subjective and the status of a particular plant species may change depending on circumstances or perspectives. Although most farmers have low tolerance of at least some weed types or species, especially in crops, other people can value them for their biodiversity or aesthetic value, and it is in this area that conflicts can arise over farming practice and wider social values. In organic farming the organic principles also lay down a set of ideals, codified under organic standards, that are intended to orientate and guide farmers as to acceptable methods of farming; weed management should also be considered in this light. Some weeds that are regarded as particularly pernicious are also subject to statutory control under the law and organic farming practices should conform to the law in these cases. Weed management will therefore be a balancing act between the negative effect of weeds and the potential benefits that they can have within a framework set by social attitudes, economic conditions, environmental and agriculture policies, as well as organic standards and the law.

Negative Effects of Weeds

The primary negative effect of weeds, and the one most often mentioned, is to reduce yields of crop plants. They do this by competing with crop plants for light, nutrients, water and space, any of which can be limiting, depending on the growing conditions. The ability of the weeds to outcompete crop plants will depend on a wide range of factors, including the weed species, the density of the weeds, the growth phase of the crop at the time at which the weeds are competing with the crop and the crop's competitive ability. These factors are all important in weed management programmes and occur throughout the text, where they are discussed in more detail.

Fig 1 Weeds in organic leeks.

Apart from the classic competition effects, some weeds can also have more direct effects on crop plants, although these cases are much rarer. Some weeds, such as dodders (*Cuscuta* spp.) and broomrapes (*Orobranche* spp.), are directly parasitic on crop plants, forming close associations with them to rob them of nutrients, water or photosynthates. Many plants, such as mugwort (*Artemisia vulgaris*) or garlic mustard (*Alliaria petiolata*), are capable of exuding chemicals that alter or suppress the growth of other plants in a process known as allelopathy. Although many 'weed' plants are suspected as having allelopathic effects, it is difficult to demonstrate this in practice as many of the effects are very similar to the end results of classic competition, as described above.

Competitive effects aside, weeds can also hinder management operations within a crop or the harvest of produce. Climbing weeds, such as bindweed (*Convolvulus arvensis*) or cleavers (*Galium aparine*), are notable for impeding harvests in cereals. Other weeds can reduce the quality of the harvested produce by, for instance, contaminating them with seed or with plant parts. They can also contaminate animal products, such as wool with burrs or other parts. Any such contamination can also impede or complicate any processing procedures, for example, by clogging machinery.

Other undesirable traits associated with weeds include being noxious or poisonous to people or livestock; for example, ragwort (*Senecio jacobaea*). They can also reduce the palatability of grass or pasture to livestock – for example, the docks (*Rumex* spp.) – and also taint livestock products – some alliums can give milk a garlicky taste. There are also some cases in which weeds can harbour pests or diseases that they can then pass on to crop plants. This can be a particular problem with soil-borne pests, like nematodes, or diseases, like clubroot (*Plasmodiophora brassicae*) in brassicas, where weeds allow the pests and diseases to effectively bridge the gap between one susceptible crop and the next within the rotation. Many species of aphids, like the black bean aphid (*Aphis fabae*), are also capable of feeding on a wide range of weed plants and then moving on to crops under the right conditions.

Weeds also have impacts on the cost of growing crops and, indirectly, on the moral of the workforce. High weeding costs are often of paramount concern in organic farming, and can be especially high in some uncompetitive vegetable crops, unless direct and timely action is taken (*see* Chapter 4 for a more detailed discussion). An understated effect of weeds is on farmer and worker morale. Many farmers equate weedy fields with poor farm management and this leads to a desire to keep fields weed free, not least to deflect criticism from neighbouring farmers who might see weedy fields as a sign of non-professionalism or as a source of weed contamination to their own farms. Failing to manage weeds in some fields and some seasons, even for perfectly acceptable reasons, can lead to feelings of resignation, losing control or simply of not being successful at farming. In addition, battling to keep fields weed free can become a daunting task to workers, and management strategies need to be employed to keep staff motivated and feeling that they are 'on top of the job'. These factors can all indirectly increase 'weeding costs' on a farm and will certainly increase the amount of time spent in staff management.

Potential Benefits of Weeds

Although the negative effects of weeds can seem intimidating, they can also have beneficial effects within a farm system and it is worth thinking about what 'services' they might be able to offer. In particular, weeds can help to maintain soil quality on organic farms in many direct and indirect ways. For instance, weeds can provide a covering for bare earth, especially in a short gap between crops, where it is not possible to sow a green manure, or even between crop rows before a crop has had the chance to grow and fill the gap. Such a covering will prevent erosion and leaching, which can be beneficial to subsequent crops and long-term soil health. There is also growing evidence that weeds help to maintain below-ground ecological cycles, by, for instance, acting as hosts for mycorrhizal fungi or other beneficial soil microorganisms. When ploughed in (and before shedding seed!) they can provide a significant quantity of soil organic matter, which can also help to drive the soil's ecology.

Weeds can aid in maintaining soil fertility and structure. Many weeds, especially perennial weeds, have extensive or deep rooting systems, which can penetrate into the subsoil and which act to break up the soil and provide good structure for crop plants. Deeply penetrating roots will also allow plants to bring up and accumulate nutrients and trace elements from deep in the soil profile, and these nutrients will be released when the plants die and decompose. Docks, for instance, are common perennial weeds that have a deep tap root and can accumulate trace elements valuable to grazing livestock and have been shown to have high levels of magnesium.

Weeds can also provide refuges for predators and parasites of pests. Many insect predators need alternate sources of nutrients or shelter, and weeds can provide these resources. There is also evidence that weeds can, in some circumstances, deter or confuse pests, such as aphids or cabbage

Fig 2 Weeds can confuse pests looking for crop plants.

root flies, by disrupting their landing and/or egg-laying behaviours. Linked to this, weeds can also be an important source of food for wildlife, especially birds. Bird populations have been declining on farmland over the last few decades and leaving weeds as a resource has been shown to help revive bird populations. Some birds will, in turn, eat crop pests.

Weeds can also be valuable indicators of growing conditions in a field; for example, of soil type, pH, compaction, nutrient levels and/or water levels. Weeds, like all plants, will tend to do better or appear in the conditions to which they are best adapted, and different species will have evolved to take advantage of different environmental opportunities. Chickweed (*Stellaria media*), for example, often appears in nutrient-rich, well-aerated soils, whilst cornflower (*Centaurea cyanus*) prefers calcareous soils and horsetail (*Equisetum arvense*) acid, poorly drained soils. Whilst there is a vast amount of information on the preferences of different weed species, much of it is not well-documented; a detailed discussion is beyond the scope of this book but can be most easily accessed through the internet (with a critical eye!).

Apart from these positive reasons for managing rather than eradicating weeds, they can be valuable in their own right or as part of the intrinsic biodiversity of the farm. For instance, many have medicinal properties (both for people and animals), and others are rare and/or regarded as aesthetically pleasing. Once again, a large amount of information on the uses and value of weeds exists and is best accessed through the internet in the first instance. Environmental schemes currently seek to encourage and augment biological diversity in farmed landscapes and weeds can surely play a part in this. Some schemes might even put a positive cash value on weeds that would need to be balanced against other income and farm costs. Organic standards (*see* below) also emphasize the importance of enhancing biodiversity and the value of weeds within a farm system.

Ecology of Weeds

From an ecological perspective, weeds are, by definition, adapted to agricultural (or man made) habitats and, as a group of plants, share many characteristics that enable them to establish in fields, out compete crop plants and to go on to reproduce and spread. The weed flora of any one farm represents a community of weeds, or mixture of weeds, that have arisen in response to the farm physical conditions (soil, weather, etc.) and management (enterprises, tillage, etc.). The weed flora will change over time in response to these factors. Like all plant communities, the agricultural plant community, which includes crop plants and weeds, is largely shaped by competition between plants for the available resources, although other chance factors can also have a significant impact on all scales; for example, bare patches of earth, localized moisture or light patches, change of land use or drought can all be important. In order to persist in the community, plants need to survive, reproduce and disperse in the face of competition and in response to chance factors, and many ecological adaptations have arisen that enable plants to exploit specific 'ecological niches' (or fulfil specific ecological roles) and be successful in this sense. Weeds are no exception.

Researchers have described a range of 'weedy traits' that can give weeds a decided ecological advantage over other plants in establishing and spreading in agricultural situations. These 'weedy traits' can include one or more of the following:

- Small seeds – weed plants typically spread 'risk' by producing large numbers of small seeds, which are easily dispersed, so that at least a few seeds will encounter the right conditions to germinate and establish.
- Vigorous and easily spread vegetative propagules (e.g. rhizomes), which can disperse over both short and longer distances.
- Rapid early growth of roots and shoots from seeds and propagules, and the ability to rapidly produce more reproductive structures, often in a much shorter time-frame than the growing crop.
- Variable or plastic phenotypic characteristics that enable them to adapt to many different field conditions and the capacity to easily regenerate or regrow if damaged.
- The ability to grow or climb above the crop canopy or otherwise aggressively compete with other plants and/or be tolerant to shading.
- The capacity to self-pollinate, but also to cross-pollinate when the opportunity occurs, and the ability to produce large numbers of seeds over a long period of time.
- The capacity to produce seeds that also have the ability to become dormant and survive in the soil for prolonged periods and germinate under differing environmental conditions and at different times. Dormancy allows seeds to survive in the weed seed bank for long periods, in which the plant itself would not necessarily survive and, although seeds will be lost from the seed bank, it can be rapidly replenished in as little as one 'weedy' season.

Thankfully not all weeds show all of these characteristics. Weeds have evolved different combinations of these traits to survive and reproduce, and this allows them to be grouped into different broad groups or types, which share similar life-cycles or strategies for surviving. In doing so they may also share similar susceptibilities to weed management techniques and operations.

Weed Types

Different weeds specialize in one or more of the above strategies for surviving and multiplying, and can be grouped according to types that follow roughly similar life-cycles. We present these briefly here, as these shared characteristics provide a basis for planning their management. All species will differ in the exact details by which they reproduce and spread (these are discussed in more detail in Chapter 5).

ANNUAL BROAD-LEAVED WEEDS

These originate almost exclusively from seed; for example, fat hen (*Chenopodium album*) and chickweed. The majority arise from seeds that have remained in the soil from previous seasons. A few will develop from seeds brought in as contaminants in crop seed, soil amendments like manure and slurry, and in organic materials, such as straw used for mulching. Some weed seeds may be blown in, while others are carried in

Fig 3 Broad-leaved weed seedlings in organic cereal.

by birds and animals or on farm machinery. In some respects they are the easiest weeds to deal with, as their life-cycle is complete in one year. The main objective should be to remove plants before they compete with the crop early in the season and later in the year before they set seed or impair the harvesting process. The cultural and mechanical weed management methods described under the weed management section can be used to effectively control annual broad-leaved weeds.

ANNUAL GRASS WEEDS

These germinate from seed and reproduce in one year; for example, annual meadow grass (*Poa annua*). These weeds can spread through contamination of crop seed, and preventative cultural control methods can be important to maintain populations at low levels and prevent seeds returning to the soil. In addition to yield effects, annual grass weeds suffer from ergot; this can result in contamination of cereal grain.

BIENNIAL WEEDS

These normally reproduce by means of seed but, unlike annuals, they rarely flower and set seed in their first year; for example, spear thistle (*Cirsium vulgare*). The first year's growth represents a vegetative phase during which the plant grows and accumulates food reserves, followed by flowering and seed production in the second year. After the plant flowers and sets seed, it normally dies. The vegetative phase might last more than one year in unfavourable conditions or, conversely, flowering may occur in the first season under favourable conditions. Management methods will be similar to those employed against annual broad-leaved weeds.

STATIONARY PERENNIALS

These are generally broad-leaved weeds that persist over many years and that multiply by seed and have short subterranean stems, often connected to stout, perennating taproots; for example, docks and dandelion (*Taraxacum officinale*). They are normally more problematic in grassland and more easily dealt with in arable or vegetable parts of a rotation. A flexible and rotational approach to management is more likely to be successful, combining a range of cultural controls to remove underground parts and direct controls such as topping to prevent seeding.

CREEPING PERENNIALS

Grass or broad-leaved weeds that extend horizontally by means of creeping vegetative organs including horizontal shoots, either above (stolons) or below (rhizomes) ground, or by thickened roots. Daughter plants are able to develop at some distance from the parent. Fragmentation by cultivation followed by regeneration may lead to a rapid increase in the number of individual plants and they can be particularly problematic in organic farming systems in both grassland – creeping thistle (*Cirsium arvense*) – and arable and vegetable crops – couch (*Elytrigia repens*). These weeds will need a flexible, rotational management approach to prevent spread, possibly including a fallow period when infestations are high.

VOLUNTEER WEEDS

Crops that become weeds are a particular problem because they can act as a green bridge that allows pests and diseases to persist, and will nullify the effect crop rotation. Where crops are grown in rotations that include a ley, clumps of pasture grasses may remain to regrow in the following crop, as can clover (*Trifolium* spp.) seedlings. The most frequent volunteers result from crops that are grown for their seeds or their vegetative propagules; for example, potatoes (*Solanum tuberosum*). Many of the novel crops grown for seed in the UK, such as borage (*Borago officinalis*) and lupin (*Lupinus alba*), also have the potential to become a volunteer problem.

Implications for Weed Management

Although not all weeds share the same weedy traits or survival strategies, they appear to be suited to a similar 'ecological niche or function' in that they are adapted to colonize and grow quickly on bare ground before other more longer lived plant species establish themselves, a habitat which modern agricultural systems reproduce on a large scale every sowing season. A sensible weed management programme would therefore rely on trying to establish any crop as quickly as possible in order to gain an advantage over the weeds, so that the crop can out compete the weeds, and then manipulate the crop so that the weeds do not get a chance to establish or are out competed. This is the basis of many of the cultural control methods discussed in later chapters, especially Chapter 2. In some parts of a rotation, and in some seasons, this will not be sufficient, and a few weeds types are likely to succeed in developing. In this case, it will be necessary to resort to more direct control measures; those available to organic farmers and growers are discussed more extensively in Chapter 3. However, it is also important to appreciate that, because of the large number of potential weed species and types (highlighted above and in Chapter 5), and their inherent adaptability, relying on one or two weed control methods, however effective in the short term, is likely to lead to longer term changes in the weed community that simply replaces one dominant species or phenotype with another. It is therefore necessary for farmers to be flexible in their approach to weed management and to vary and rotate the management methods used.

Organic Standards

Apart from the practical considerations of weed ecology, organic farmers also subscribe to organic principles, a set of values codified as standards by certification bodies. Although there are many certification bodies active in the UK and EU, most such schemes are also legally underpinned by EU law (EC Council Regulation 2092/91), which defines the minimum requirements for organic standards. Although there are a large number of different certification bodies across the EU, many aim to incorporate the IFOAM organic principles within their standards and they have similar attitudes regarding weeds and acceptable weed management practices.

However formulated, organic principles usually stress the importance of biodiversity to, and within, farming systems. The IFOAM principal of ecology states that 'Organic Agriculture should be based on living ecological systems and cycles, work with them, emulate them and help sustain them' and on this basis weeds should, therefore, be regarded as integral to the farm system. Most, if not all, certifying bodies recognize that ecological diversity is integral to organic farming, and wildlife habitats should be managed as such and should comply with any ongoing conservation planning, including any 'weeds' that fall under such conservation management plans.

More ecologically intrusive or disruptive methods of weed management are not usually allowed in organic systems. For instance, steam pasteurization or soil sterilization is not allowed for weed control because of the likely negative impacts on soil microorganisms. In addition, most certifying bodies do not allow the use of any synthetic or organic 'herbicides' or weed killers because of the secondary effects on biodiversity and soil ecology.

In many respects the requirements for weed management, under the standards, do not differ from those dictated by the ecology of weeds and should generally be undertaken from a whole-system point of view. Planning for weed management should also incorporate good rotation design, adequate manure management, well-timed soil cultivation and good farm hygiene. Many of these methods are described in detail in the subsequent chapters.

Statutory Regulations

In addition to the guidance provided by organic standards, some weeds are also subject to statutory control (e.g. under the Weeds Act 1959 in the UK). Defra (through the Secretary of State) may serve an enforcement notice on landowners who have an infestation of 'injurious weeds', requiring them to take action to prevent their spread. The Weeds Act specifies five injurious weeds: common ragwort, spear thistle, creeping thistle, broad-leaved dock (*Rumex obtusifolius*) and curled dock (*Rumex crispus*). More recently, the Ragwort Control Act amended the Weeds Act and was intended to promote the more efficient control of common ragwort, which is the only weed specified in the original Act that poses a risk to animal health. The Ragwort Control Act came into force in February 2004 and Defra has produced a code of practice on how to manage and dispose of ragwort, available on their website. A complaint form is also available from the website for anyone wishing to complain about an infestation of injurious weeds, as specified in the Weeds Act 1959.

WEED MANAGEMENT ON ORGANIC FARMS

Weed management is generally understood to be a process that manipulates the crop and cropping practices to the advantage of the crop and the disadvantage of the weeds. The aim of organic weed management is to maintain a low and tolerable level of weed infestation within crops and on the farm. It should be appreciated that eradication of any particular weed is likely to be unfeasible and, even if achievable, likely to be economically unviable. Even if a particular weed species were removed, it would simply be replaced by another weed species. Weeds are therefore an integral part of all organic farm systems and, from a farmer's perspective, need managing to reduce their negative impacts and, once this is taken care of, to enhance their positive impacts.

Effective weed management in organic farming systems will depend on a large number of contingent factors, some of which the farmer will be able to control but a large number of which will be outside their direct control. All farms, but especially organic farms, are complex systems with many different rotations, enterprises, market systems and farmer values. Although weed management technologies and techniques underlie or form the rationale for many farm operations, at the end of the day, they are just one aspect of the farm ecology and socio-economics (but an important one!). Weed management decisions, therefore, need to be made within a complex framework that depends on economic, social and environmental circumstances at any given time. In fact, a deep understanding of any specific situation or decision depends on many factors at field level (such as field history, crop sequence and weather), at farm level (such as farm size, enterprises and location) and at regional level (such as markets, policy and environment).

As a consequence of this complexity, answers to weed management questions, such as how to control docks, are likely to be very situation-specific. In general terms, no farmers (or advisors or researchers) think that they have all the answers to weed management problems. Successful organic weed management is likely to require an integrated systems approach, and a range of methods and techniques will work best over a period of time. In practical terms, such an integrated approach will necessarily be a mixture of longer term preventative planning, such as rotations and crop variety choice, combined with short-term reactive or direct measures, such as harrowing and flame weeding. Unlike conventional systems that rely on herbicides to 'backstop' weed control, organic systems cannot be understood solely in terms of one or two components; all the other parts that make up the complete ecosystem need to be taken into account and allowed to contribute to weed management. Some of these interactions can be very complex. For example, soil fertility can influence crop–weed interactions from the effect of nitrate on the timing and extent of weed-seed germination, through to the differences nutrient level and composition can make to the balance of plant competition. Both weed emergence and nutrient release are also closely associated with soil disturbance. However, soil cultivation is also an important component of direct weed control, as well as of the seedbed preparations for crop establishment.

These types of circular system effects can be sometimes difficult to grasp, which has led to the conclusion that successful organic weed management is also as much about learning for improving complex situations (*see also* Chapter 7), as about actual weeding techniques. Farmers and growers should be, therefore, constantly seeking new knowledge on weeds to incorporate into their weed management strategies, and continually adapting and trying out different weed management techniques, in other words to engaging in 'learning' and 'research' in its widest meaning. Farmers seeking information on weeds can use the results of detailed scientific trials and botanical observations, but these should be regarded as two elements in a large pool of knowledge from which they can draw on to make weed management decisions. For example, other farmers and machinery manufacturers also have a great deal of experience and knowledge about weed management and, in many cases, this more 'practical' knowledge can be invaluable. Ultimately, farmers should also take the opportunity to undertake informal experimentation with weed management techniques and methods in their own fields (trialling to see 'what works', learning from chance observations or events), as this is likely to lead to weed management strategies tailored to their specific circumstances. In fact, successful weed management strategies are likely to be the outcome of this experiential learning process, usually in the form of a toolbox of practical

Fig 4 Learning about weeds from different sources is important.

weed management methods, applicable in specific situations and tailored to a farm situation. Although these methods need not be 'formally' evaluated (in a statistical sense) they can, nevertheless, be part of a systematic approach to building knowledge to effect specific weed management situations.

In summary, all farmers need to adapt their farm systems, within the resources available to them, to manage weeds at some level. In doing this, weed management practices should aim to benefit the crop over the weeds, balance the costs and benefits of weeds, maintain organic principles (such as sustainability and biodiversity), plan long-term, but remain flexible, and aim to maintain low and tolerable levels of weed infestation. This is likely to make organic weed management more complex (and costly!) and growers should be constantly seeking new knowledge and looking to use, and incorporate it, into their weed management strategies.

In writing this book we have attempted to provide a means by which farmers and growers can do this. We have done this by asking ourselves a series of questions that any farmer or grower needs to ask before taking any weed management decision and which we have attempted to answer in writing this book. We provide these below as a pointer to our thinking.

Is weed control needed? What are the weeds (*see* Chapter 5)? Will marketable yield be affected by the weeds? What other negative effects are the weeds having? What are the long-term consequences of infestation? Are there any benefits associated with the weeds? What is the likely balance between costs and benefits (*see* Chapter 4)?

If control is needed:

Which method of control? Can longer term cultural and preventative measures be used (*see* Chapter 2)? Is direct physical control needed (*see* Chapter 3)? Can biological control help (*see* Chapter 2)?

If direct control is needed (*see* Chapter 3):

Where is control needed? How closely can the weeds be targeted? Is a broad-spectrum approach necessary, in other words weeding across the entire field area? Or inter-row weeding? Or intra-row weeding? Will a patch weeding approach suffice?

When is control needed? When is the best time to weed? When are the 'critical periods' of competition? How many weeding operations are needed?

Chapter 2

Preventative and Cultural Weed Management

Weed management can be defined as the use of any technique that favours the crop plant over the weeds. Organic farmers and growers should take a preventative approach to weed management; that is, try and side-step weed problems before they arise or reach a point where more drastic direct action is necessary (*see* Chapter 3). Crop planning is the cornerstone of this preventative approach and rotations can be designed to positively influence weed populations at the same time as being integral to the whole-farm management strategy. Within the planning and rotational approach, a whole range of options are available for farmers and growers; these are discussed in this chapter. However, it should be stressed that flexibility is also important and that sufficient leeway should be built into any planning process that allows for dealing with any unforeseen circumstances, and to take any opportune moments to augment weed control on a practical day-to-day level.

CROP ROTATION

The planned rotation of crop types is a central concept when designing or modifying organic farming systems, although it may be curtailed in more extensive systems that include permanent pasture and in perennial crops such as fruit. The length of the rotation, the choice and the sequence of crops will usually depend upon individual circumstances that include factors like soil type, rainfall, topography and enterprises, as well as the potential weed burden. Traditional rotations have included legume crops that serve to improve soil fertility and grass or legume 'sod' crops or leys to maintain soil organic matter.

Crop rotation has important implications for weed management and provides the main opportunity for indirect or preventative weed control. No one rotation can be recommended but, ideally, in terms of weed control, the aim should be to achieve diversity, whenever and wherever possible. The underlying principle of a preventative approach is to produce a constantly changing environment to which no single weed species can adapt and become dominant and unmanageable. Success depends on the use of crop sequences that create varying patterns of resource competition, allelopathic interference and soil disturbance. In practical terms this means as diverse and as long a rotation as possible, consistent with the farm system and which prevents the weeds returning seeds to the soil seedbank, as well as preventing the spread of other propagating structures, such as rhizomes.

Weeds tend to flourish in crops with requirements similar to their own. In monocultures this results in the development of characteristic weed floras. Rotations aim to disrupt this and provide an unstable and changing environment that prevents the proliferation of particular weed species. In horticultural systems this is likely to be a sequential cropping pattern, where many different short-term crops follow each other in quick succession, both within a season and between seasons, perhaps punctuated by shorter periods of cover crops between cash crops, and including longer periods of fertility building crops during one or two seasons. In contrast, a typical arable rotation may extend over several years with an annual change of a limited number of crop types, but still interspersed with fertility building leys. Introducing livestock into a farm system will help to diversify the rotation and can be a positive bonus for weed management, although it is also not without risk in spreading weeds. Grass systems may include areas of permanent pasture, which bring their own weed management challenges. In all systems, the inclusion of cover crops, intercrops and green manures can increase crop diversity in a rotation and help to suppress weeds. The implications of many of these choices for weed management are discussed in more detail below.

Crop Choice and Sequence

A general aim should be to take advantage of the different stratagems involved in growing a range of crops with diverse growth habits and production needs, so that a whole range of differing factors can be bought into play. At its most basic, rotations should include different crop types (for example, cereals, potatoes and leafy vegetables), which will naturally lead to a diverse cropping environment. Exchanging different crop types will alternate between those that require a longer growing season and others that mature more quickly. It might also be possible to alternate between annual and longer term (biennial or perennial) crops. Open crops may follow or precede those that develop a dense leaf-canopy, the latter being better at suppressing weeds. Typical so-called 'cleaning crops' include turnip (*Brassica rapa* var. rapa), sugar beet (*Beta vulgaris*) and potato. Among the cereals, oats (*Avena sativa*) and winter rye (*Secale cereale*) are the most competitive, followed by triticale (*Triticale hexaploide*). A competitive cereal like rye may be grown as a short-duration 'smother' or 'cover' crop in the rotation, and allelopathy may also play a part in weed suppression.

The time of planting will determine the species composition of the weed flora likely to emerge following seedbed preparation. Spring emerging weeds will predominate in spring sown crops, autumn emerging weeds in autumn sown crops. Regular alternations between spring and autumn sown crops can therefore create a range of opportunities for controlling weeds at different times, and will prevent any one weed flora adapting to the rotation. The timing of sowing can be adjusted to give the crop a 'head start' or advantage over weeds; for instance, by delaying sowing until warmer conditions prevail, and larger seeded crop types can develop quickly to out compete weeds. Stimulating premature weed emergence before sowing gives an opportunity for weed control prior to sowing the crop. The timing of harvesting can also affect the types and stages of development of weeds present in fields with relatively early harvests giving less opportunity for weeds to develop and seed.

The inclusion of different crops in a rotation will also allow a greater variation in the range of cultivations and non-chemical weeding methods that can be used, as compared to those that are possible with a monoculture. For instance, the inclusion of a row crop can provide an opportunity to reduce the soil seedbank, and hence potential weed numbers in future crops, by allowing the use of regular cultivations that kill emerged annual weeds and encourage the germination of others. Cultivations may also reduce the problem of perennial weeds by disrupting their growth until the crop canopy is able to smother further regeneration.

Fertility Building Leys

Grassland and grass–clover leys are an important part of any organic farming system in the UK and temperate zones. Where a long grass break does not form part of the rotation, weed problems are likely to be more severe – especially when less competitive crops are grown. On livestock farms, grassland forms the basis of the production process; whilst in arable and horticultural systems, leys are used primarily for maintaining or restoring soil fertility. A ley is generally seeded with a mixture of perennial species, and is maintained for several years before the land is ploughed and returned to annual cropping. The grass may be managed as a short, medium or long-term crop, and this will determine the composition of the desirable sward species and the nature of the associated weeds. The seed mixture for a ley can include a relatively simple mixture of grasses – usually perennial ryegrass (*Lolium perenne*) – and legumes – normally red or white clover – or may be more complex and contain a range 'herbs' such as yarrow (*Achillea millefolium*), which in other situations would be considered a weed. Many consider that such species add to the nutrient value of the grass and benefit both the stock and the land. Burnet (*Sanguisorba minor*) and yarrow, for example, are thought to help control diarrhoea in sheep. Even species such as dandelion that invade later, are seen by some as beneficial. The composition of the sward will change with time and can be modified to some extent by the management strategy adopted.

A grass or grass–clover ley provides a completely different habitat to that of a field crop and may be used to reduce or eliminate particular weed species. If managed well, the ley period can act as a weed suppressing phase. Rotations that include grass leys have been shown to be beneficial in reducing weed seed numbers compared with rotations that do not have a grass phase. For instance, leys were a traditional way to deal with land infested with wild oat (*Avena fatua*) although it does not eliminate this weed completely. The choice of fertility building crop and ensuring good establishment are both important. The length of the fertility building period within the rotation will also have an impact on the weed population. A greater proportion of ley in the rotation usually results in lower seed numbers in the weed seedbank in comparison with annual crops because a percentage of the weed seeds in the soil beneath the ley will lose viability each year and there should be no further addition of fresh seed in a well managed ley.

Although few long-term studies on enhancing the effectiveness of leys in managing weeds have been carried out, trials suggest that there is little advantage in leaving leys down longer than three years. The species composition, and the mowing and grazing regimes of the ley, are important in the dynamics of above-ground vegetation but have a limited effect on the below-ground seedbank. Cutting and topping are important for weed management in pasture, grass and leys. Ley management should include topping at intervals during the summer to a height of around 10–15cm (4–6in). Ideally, in such fertility building leys, the sward should not be allowed to get higher than 40cm (16in) or at most knee height. If the vegetation gets higher than this, then topping will create a mat of vegetation that will act like a mulch, which can create dead spots in the ley where clover may be excluded by the more vigorous grasses, or which weeds may colonize. Topping the ley regularly will also ensure that tall weeds that may have germinated will not be able set seed. Topping can also be used as a remedial measure in vegetable or other crops to prevent weeds from seeding. In grassland systems, temporary leys provide an opportunity to control perennial weeds during the cultivations between ploughing and reseeding.

Management of the weeds at the time of ley establishment is critical, as is the method of ending the ley, to avoid a flush of weeds due to the release of seed dormancy by cultivation. Establishment of leys can be easier in the autumn period than in the spring because sowing in spring coincides with the main spring flush of weeds. The seedbed needs to be well prepared, and good contact made between the seed and, ideally, the moist soil to achieve good germination that allows the ley to establish and out compete any weeds. Volunteer ley species may also be a problem in subsequent crops, both from seed and from regenerating pieces of sod.

In deciding which crops to place after a ley there may be a conflict between taking advantage of the enhanced fertility available as compared to taking advantage of the likely reduced weed burden. Nevertheless, it is often advantageous to plant poorly competing crops directly after a perennial ley because, by the third cropping year after the ley, there is likely to be twice as much weed emergence as compared to the first.

Break Crops, Cover Crops and Living Mulches

Introducing cover or break crops into a rotation can also be important for weed management. Break crops are so-called because they 'break' the cycle of normal cropping and provide diversity in rotations. This change of cropping may have several benefits: improved pest and disease control, more efficient nutrient use, higher economic returns, as well as improved weed management. Potatoes, for example, could be used as a break crop in an arable rotation as they are a broad-leaved crop with a different morphology, growing system and associated weed flora compared with a cereal crop. Potatoes also allow the use of different weed control methods providing an opportunity to cultivate the soil with an alternative range of implements, as compared to the cereal. More exotic organic break crops could include sugar beet, hemp (*Cannabis sativa*), linola (*Linum usitatissimum*) and lupin (*Lupinus* spp.).

Cover crops have the primary purpose of covering the soil, although they can also be grown between the rows or plants of a cash crop, and in this situation are often referred to as living mulches. They are principally used for preventing erosion and for nutrient management, often by absorbing nitrates from the soil to prevent them leaching as 'catch crops' or as green manures, which provide ample organic matter when ploughed in. Legume cover crops also fix nitrogen. The inclusion of cover crops in the rotation, at a time when land might otherwise lie uncropped, will also

Fig 5 Vetch can be used as a weed-suppressing cover crop.

serve to suppress weed development. Cover crops also tend to be quick germinating and dense, so that they suppress weed emergence. It is essential to get good establishment of a dense crop to prevent weed establishment, and for this reason the ground should be adequately prepared. Some cover crops may also exhibit allelopathic properties, in other words they exude chemicals that can have an inhibitory effect on surrounding plants, including weeds. For example, when perennial ryegrass has been incorporated and is breaking down, it has been shown to reduce the emergence of weed seedlings in a subsequent crop. In addition, the phytotoxic effect of ryegrass lasted longer than that of clover. Residues of autumn sown rye that was killed in the spring, has been observed to reduce weed biomass by over 60 per cent because of its allelopathic properties (over and above its physical effects).

In horticultural systems, cover crops can be an essential component of the rotation. Some stockless vegetable production systems rely on short bursts of green manures, as well as leys, for fertility, in which case, manipulating the cover crops for weed management can provide additional opportunities to control problem weeds. Cover crops in such systems can be managed in a variety of ways to suppress weed populations. They may be sown in the autumn and killed off before vegetable crops are seeded in spring, although destruction through incorporation greatly reduces any weed control benefits, as fresh seed is bought to the surface by cultivation. Using frost sensitive cover crops eliminates the need for destruction in spring, but earlier establishment is needed to obtain good ground cover before the first frosts. Green manures can be cut and left on the surface as mulches, which can suppress weed emergence and growth. Although allelopathy may be involved, other factors, such as light transmittance, soil temperature and soil moisture under the residue, are also important. The plant residues also often provide a protective habitat for seed predators and this may also help to reduce weed seed numbers. Mulches are discussed in more detail in Chapter 3.

The use of cover crops and living mulches also may bring some disadvantages: they may affect the seedbed preparation for following crops; they can promote a high carbon to nitrogen ratio in the soil, which 'locks up nitrogen'; they can act as a source of disease infection to crops under some circumstances; and can provide a refuge for slugs (various species). In this case, the balance between being a cover crop and a potential 'weed' can be quite fine. Lucerne or alfalfa (*Medicago sativa*) residues exhibit a strong phytotoxic effect on a range of crops (including itself) but improve the growth of some vegetable crops. Decomposing cover crop residues may release allelochemicals that inhibit the germination and development of weed seeds. Unfortunately, drilled, small-seeded crops may also be adversely affected. Residues of ryegrass (*Lolium rigidum*) and subterranean clover (*Trifolium subterraneum*) cover crops have been shown to reduce seedling growth of lettuce (*Lactuca sativa*), broccoli (*Brassica oleracea* var. italica) and tomato (*Lycopersicon esculentum*).

Inter-cropping and Undersowing

Inter-cropping is widely practised on a global scale and an enormous variety of inter-cropping systems have been developed. In successful inter-crops, weed suppression is usually superior to that of either of the component crops when grown alone. Both component crops may be taken to yield or one may be there as a living mulch to improve weed control or provide some other benefit like nitrogen fixation. Crop density, crop diversity, crop spatial arrangement, choice of crop species and cultivar will all affect weed growth in inter-cropping systems.

Although inter-cropping two cash crops has not been widely adopted as a practice in temperate organic production systems, there have been investigations into mixed cropping of organic winter wheat (*Triticum aestivum*) and field beans (*Vicia fabae*), which reduced weed growth and gave better yields than sole cropping. Inter-cropping leeks (*Allium porrum*) and celery (*Apium graveolens*) has also been shown to increase weed suppressing ability, and reduce reproductive capacity of late emerging groundsel (*Senecio vulgaris*). A large range of other crop combinations are also likely to give weed control benefits but most studies indicate that improved weed control alone is unlikely to justify their use and there must be other obvious benefits if the change in cropping practice is to prove economic.

Undersowing can be seen as a form of inter-cropping or as a practice in its own right. Weed emergence can be reduced if a quick growing, dense layer of vegetation is established under a crop. If cash crops are undersown with legumes, then there can be advantages over and above weed suppression, primarily additional fertility building effects, which might lead to reduced need for lengthy isolated fertility periods. This needs to be balanced against likely loss of crop yield when economic returns are calculated. However, despite a fair amount of research effort with vegetables, undersowing has only been widely adopted for use in cereals in the UK, as it greatly aids the establishment of leys after cereal crops. In this case, the undersown crop can help suppress weeds within the cereal crop but, more importantly, the prior establishment of the ley denies weeds an

Fig 6 Undersown sweet corn.

opportunity to gain a foothold in the new ley. Although there are examples of more novel vegetable production systems using undersown legumes, they have not been widely adopted, often because of the depressive effects on the marketable vegetable yield due to the difficulty in timing the sowing of the undercrop to avoid competition, especially in low growing vegetables. Undersowing is easier in taller crops such as sweet corn (*Zea mays*) or climbing beans, for example some varieties of French beans (*Phaseolus vulgaris*) or runner beans (*Phaseolus coccineus*).

Fallowing

A fallow period allows weeds to germinate and develop in the absence of the crop, so that they can be more easily dealt with, normally by repeated mechanical cultivations (*see* Chapter 3). It is not usually desirable to plan for a fallow period in a rotation as it is obviously non-productive and can be costly, both directly and indirectly, due to damage to soil structure and fertility. It may, however, be necessary if weeds, especially perennial weeds, cannot be controlled during cropping or fertility building. Tillage without a crop for a whole season is sometimes referred to as a black fallow. It may not be necessary to stop cropping for a whole year, but instead to employ a bastard fallow, that is, no crop for part of the year. Fallowing the land for part of the growing season, as a bastard or summer fallow, can be as effective as a full fallow; it is more suitable for lighter land and can be fitted into most rotations. It is often more effective during the summer when frequent cultivations can take place and when the drier periods allow for root desiccation.

Fallowing has been shown to reduce perennial weeds within a rotation. The aim is to kill the vegetative organs of the weeds by mechanical damage and desiccation. For a full or bare fallow, heavy land is ploughed in April to give the weeds time to start into growth. It is cultivated or cross-ploughed ten to fourteen days later to produce a cloddy tilth. The soil is cultivated or ploughed at frequent intervals to move the clods around and dry them out. By August the clods should have broken down and the soil is left to allow the weed seeds to germinate. In September/October the weeds are ploughed in and the land prepared for autumn cropping. If a cereal is to follow the fallow, wheat bulb fly may be a problem because it lays eggs on bare ground in July. This can be overcome by sowing a green manure such as mustard (*Sinapis alba*) to cover the land during this period. In contrast, ploughing a summer fallow can begin in June/July, allowing time for an early crop to precede it. One method is to cultivate the soil progressively deeper over time, exposing underground plant parts to desiccation at the soil surface; consequently, dry weather conditions are essential. There is also an opportunity for birds to feed on wireworms exposed during soil disturbance.

Fallowing as a technique is more useful in plough dominated systems rather than grassland management. Perennial weeds are usually more of a problem in grassland and have a capacity to regenerate and spread vegetatively as well as by seed. In theory, cultivation could keep them in check, but much more research is needed into machinery, time, frequency and height of cutting, removal methods and also prevention techniques. A bastard fallow is often used after a ley to reduce perennial weeds before sowing a winter cereal.

Although there is the benefit of reduced weed control costs in subsequent crops after an effective fallow, the economics of taking land out of production for a full year, together with undesirable effects on the soil and the environment, make the use of a bare fallow of dubious value as a key method for weed control in the organic system. There is no financial return during the fallow period, while labour costs accumulate during the fallowing operations. As an alternative to fallowing, break crops such as potatoes and turnips allow repeated hoeing for weed control (but are not so suited to heavy land). A similar effect to that of fallowing can be achieved by growing rapidly developing crops like radish (*Raphanus sativus*) that are harvested before the onset of weed competition. The short interval between crop establishment and harvesting in this crop encourages weed seed germination but does not allow the weeds time to set seed or reproduce vegetatively.

CROP MANAGEMENT

Within the rotation, the management of individual crops will obviously have important ramifications on weed control. There are a large number of factors that can be potentially manipulated during the cropping phase to suppress weed development, starting from the choice of variety through crop establishment to crop management and harvest. Many of these factors are discussed below.

Variety Choice and Seed Quality

Organic systems need varieties that actively suppress weeds, and farmers and growers should try and choose varieties that display at least some weed suppressing characteristics. Quick germination and establishment, rapid early and vigourous growth, and the ability to rapidly cover the soil and shade it (prostrate or tall varieties) in order to out-compete weeds at an early a stage in the crop cycle, are all desirable traits that will potentially shade out weeds. Crop varieties with a larger seed size have also been shown to exhibit greater initial vigour of emergence and growth, which may subsequently provide extra competitive ability.

Spatial distribution of the canopy foliage and rooting system will also be important and those varieties that tend to cover the ground area, and/or root extensively, will tend to out-compete weeds, especially if given a head start. Varieties are well known to differ in architecture and rooting ability, and whilst those that out-compete weeds are preferred, it should also be borne in mind that those with erect foliage, or that can tolerate some degree of mechanical weeding, are also likely to be useful. Some varieties seem to exude allelochemicals that work to suppress weed development but this is not well understood.

If there is a choice available, then the most vigorous species should be selected, as these will be more likely to out-compete weeds and suppress their development. The trend in organic cereal production has been to grow the taller stemmed varieties for their weed suppressing ability. Some farmers have stayed with the shorter stemmed varieties and employed a weed topper/cutter, which will remove and, ideally, collect weed seedheads, so long as there is a difference in height between crop and weed. In grassland systems the choice of variety may be dominated by forage value but, if there is opportunity, the most vigorous species should be selected, as these will determine the productivity of the whole ley period. In vegetable systems there are a large number of varieties available within most crop-types, and weed suppression will only be one characteristic among many that growers will have to juggle when choosing varieties.

High seed quality can be important, both for good germination and for being free from contamination. Well stored, disease free seed is more likely to germinate quickly and establish, giving the crop a head start over any weeds. It is also important that crop seed is free from contamination by weed seeds. In some countries, tolerance levels are defined for certified seed and there are often different tolerances for different crops, with large-seeded crops, such as cereals, generally having lower tolerances than small-seeded crops, such as clover. It is also important, when saving seed, that it is taken from weed free crops and, ideally, professionally cleaned so as not to spread weeds to any uncontaminated fields. If contamination is suspected, commercial services can provide an analysis.

Seed Rate and Crop Spacing

In drilled or transplanted crops, the proximity of the crop plants to each other will determine the competitiveness of the plant stand as a whole.

Fig 7 Cereal drilled in rows to facilitate mechanical weeding.

Spatial distribution of the canopy foliage and rooting system will be important for weed suppression; the general principle being that the greater the amount of space taken up by the crop, the less space there is available for the weeds to invade. However, it should be borne in mind that closely spaced crop plants compete with each other and that it is also expedient to allow sufficient space between plants to allow for efficient mechanical weeding, should weeds develop and threaten the crop. There is, therefore, usually an optimum crop density, which balances crop competitiveness against weeds with yield. In order to achieve uniform and closer spacing, higher seed rates tend to be chosen for organic as compared to conventional crops. Apart from increasing the crop competitiveness, this also allows for potentially lower germination rates and loss of the crop to mechanical weeders.

Row spacing, pattern and direction can all influence weed incidence, principally by altering the amount of shade and the microclimate within the crop canopy, although it is difficult to draw general conclusions as environmental factors, like soil nutrient status, are also likely to be important. There has, for instance, been a lot of work in cereals on row spacing, pattern, direction of sowing and seed rates (typically 10 per cent higher in organic cereals) to try and augment weed suppression (among other factors). The results have tended to be varied, with a lot of interaction between different factors, depending on the exact situation. For example, with narrow row widths an E–W sowing was favourable, whereas with wide row widths, a N–S sowing showed better response. In contrast, in vegetable row crops, market size specifications are often the main driver in determining the crop spacing, rather than weed suppression potential. Farmers and growers will need to experiment to determine the seed rates and crop patterns that are best suited to their rotations and cropping requirements.

Soil and Nutrients

Ecologically, different weeds have evolved to exploit slightly different niches, especially with regard to soil and soil nutrients, and this is likely to be one of the prime determinants of the background weed flora. Soil condition and nutrient availability can also potentially have a great influence on the weed flora and on the outcome of competition between the crop and weeds. Some weeds, such as corn chamomile (*Anthemis arvensis*) or corn spurrey (*Spergula arvensis*), are typically favoured in acid conditions and liming may help to reduce their incidence. Similarly, rushes (*Juncus* spp.) and horsetail are favoured in wet conditions, and management can be aided by promoting a more loosely structured soil, by sub-soiling or by drainage.

Apart from soil conditions, competition between weeds and a crop can be influenced by nutrient availability. Weeds could be expected to quickly respond to the presence of soluble nitrogen, as they are adapted to exploit 'bare ground', a situation in which competition for soluble nitrates could be expected to be low, and in this situation abundance of nitrate might even be a cue for them to germinate and develop. Indeed, it has been noted that the slow release of nitrogen from legume residues does not appear to encourage weed growth as much as nitrogen from applications of fertilizer, an observation which will tend to benefit organic farmers and growers. The timing of nitrogen availability is also a factor affecting crop–weed competition, and slow or delayed release is likely to favour the crop over the weed, especially as weeds generally have smaller seeds and will be more dependent on soil nutrients after germination.

Similarly, weeds can be favoured by the presence of other macro- or micro-nutrients. For instance docks have been noted, in some studies, to prefer low potassium levels, although other studies have not shown this to be the case. Creeping thistle has been observed to be worse in older grassland soils with low phosphate or high potassium levels. In all such studies it is difficult to determine cause and effect, and it is not clear to what extent reversing a nutrient deficiency will also help to mitigate a specific weed problem.

Notwithstanding this, it is clear that the outcome of competition between the crop and any weeds is likely to depend on a range of factors that interact in a complex manner, of which soil nutrient availability is only one. A general principle would be to try and provide the crop with optimum nutrient availability at an appropriate time that allows it to properly develop and to out compete the weeds, without allowing a surplus of nutrients to accumulate that could be used by weeds to promote their own growth. Apart from fertility building leys, most organic farms are likely to use manures, composts and/or slurry as part of their nutrient management programme, and some thought should be given to placing these at a position in which the crop is more likely to benefit than weeds; for example, by injecting slurry below the surface. In any case, all such inputs should be free from weed seeds and this is discussed in more detail under farm hygiene below.

Crop Establishment

The ability of the crop to get off to a good start ahead of the weed flora is critical. Good soil management practices (*see* Tillage below) are important to provide the best possible seedbed in which to plant a crop and in providing adequate nutrition for the developing crop (*see* above). A compacted soil or poorly prepared seedbed can result in poor crop establishment and subsequent weed invasion. In some systems, crop emergence can be aided by the use of primed seed, which germinates quickly and which can provide the crop with a head start. Delaying sowing until the soil is warmer can be advantageous for some crops such as parsnips (*Pastinaca sativa*), as it can speed up crop germination, emergence and establishment. However, this needs to be balanced against other factors like marketing or lack of moisture, which may damage crop yield.

Transplanting is a popular technique in organic horticultural systems, whereby an already established plant is planted into a freshly prepared weed free seedbed. Bare-rooted transplants can be raised on holdings or modular plants raised or bought in and then planted out in the field. This also has the benefit of allowing accurate spacing, that is by not having to rely on germination that can sometimes lead to uneven establishment with subsequent opportunities for weed invasion. Transplanting also accentuates

Fig 8 Transplanting leeks into a clean seedbed gives them a head start over weeds.

the difference in size between crop and weed, which can be vital for efficient mechanical weeding at later stages.

Pasture Management

Pasture crops are more permanent than arable or horticultural crops and will need a different approach once established. Weed management is generally carried out by cutting and topping weeds or grazing, which in turn will have an impact on the type of weed flora in a field and can be invaluable in preventing return of weed seed to the soil seedbank. However, the basis for good management is maintaining the condition of the sward by cultural means. In particular, reducing weed intrusions by chain harrowing in spring and topping regularly during the growing season can be important in preventing weeds becoming intrusive. Mowing during the seeding year must be carefully judged and close cutting, which can allow weeds to establish, avoided. In addition, spring sown stands should be cut no later than mid-August to allow recovery before winter, summer sowings should be left unmown until November, and undersown lucerne should be left to grow into the winter. Where companion grasses are growing strongly, light winter grazing may be desirable. In grazed pasture, weeds that are not eaten by livestock will need to be topped to prevent seed shed.

Cutting for hay or silage will also have an impact on the weed flora. The established crop may be cut up to four times per year starting in mid-May. The crop is quickly weakened by defoliation, for instance by grazing or cutting at too young a stage, especially in spring or autumn. Before entering the dormant (over-wintering) stage the crop must also be allowed to make sufficient growth to replenish the food reserves in the root. Silage tends to be cut early in the season when the sward is young and fresh, whilst hay is cut at a later stage. There can be both advantages and disadvantages associated with the timing of cutting, depending on the weed flora and the ultimate requirements of the system. Cutting late may allow weeds in the pasture to grow to maturity and set seed, which, in turn, may contaminate the hay and remain viable when passed through livestock. Dock seeds should not survive low pH silage; however, they will survive in a later cut of hay. Mature seed may also shed on the ley surface and find opportunities to germinate *in situ* or be transported by livestock to other locations. In contrast, cutting early for silage in fields, with, for example, an infestation of creeping thistle, may encourage the spread and growth of this weed. Hence, a balance needs to be established between the requirements of the farming system and weed control.

TILLAGE

Soil cultivation, or tillage in its various forms, has long been the mainstay of weed control and is the most effective way to reduce the weed seedbank. Seeds are encouraged to germinate and then the soil is cultivated mechanically to kill off the seedlings and plants. In the UK, the mouldboard plough is the traditional implement for burying weeds and crop residues, and as ground preparation for establishing a new crop. One piece of research showed that the annual loss of seeds from a natural soil weed seedbank (with no addition of fresh seed) was 22 per cent with no cultivation. When the soil was cultivated twice a year, the annual loss was 30 per cent; and when cultivated four times, it was 36 per cent. However, it is not just the cultivations associated with the post-harvest incorporation of crop and weed residue that have weed control benefits. The method, depth, timing and frequency of cultivations may influence the composition, density and long-term persistence of the weed population.

Tillage is often divided into three types: primary, secondary and tertiary. We have provided an outline of the main tillage types and their effect on weed management below, whilst recognizing that there are many operations that do not easily fall into one or other of these categories or, indeed, span them all. In addition, and in common with any other system, there may be conflicts in the ultimate goal of tillage operations. Finer seedbeds produce more weed seedlings but a smooth surface makes direct weed control easier. Larger clods of soil produce fewer weed seedlings but the rough surface gives the weeds that do emerge protection against direct weeding operations. Excessive cultivation can also harm soil structure leading to capping of the soil surface and, in the longer term, to loss from erosion. Under reduced tillage there is better control of soil erosion, conservation of soil moisture and more efficient use of fossil fuel. However, not all soils are suitable for reduced tillage and weed control may be poor, especially in organic systems.

Primary Tillage

Primary tillage is the principal cultivation method used prior to crop establishment. The main choice is between plough or non-plough (reduced or minimal cultivation, conservation tillage, no-till, direct-drilling and so on) systems of soil management. In addition, it is also possible to distinguish between inversion and non-inversion tillage. A wide range of machinery is available for primary tillage, and some combinations can even work a stubble down to a seedbed in a single pass.

There will be advantages and disadvantages inherent in different tillage methods and the use of different equipment (for example, mouldboard ploughs, disc ploughs, chisel ploughs, power harrows), which will move the soil to different depths and/or mix it in different ways. In addition to the effect on weeds and weed control, the use of different cultivations may also have implications for disease and pest control, and for nutrient cycling. Excessive cultivation can harm soil structure (depending on soil condition) leading to capping of the soil surface and, in the longer term, to loss from erosion. Running machinery on the land will compact the soil and destroy larger pores in the surface soil, and can leave an uneven surface that allows puddles to form in any depressions. A plough pan may be built up and need dealing with. Tillage can also affect the fauna in the soil. Some species of earthworm can withstand light tillage but intense tillage can reduce earthworm populations due to reductions in food supply, damage to soil burrows and direct injury. Such changes in soil structure and fauna may, in turn, affect the weed flora in a field.

Most organic farmers and growers find it impossible to routinely dispense with the plough or some form of inversion tillage. Ploughing is useful as a method of burying weed seeds and weeds below the depth from which they are capable of germinating and growing. This is particularly

Fig 9 Inversion ploughing buries weed seeds.

useful for small seeded and annual grass weeds, which are often short-lived and may survive being shallowly buried. However, it is sometimes said that ploughing buries a weed problem rather than solves it. For example, although it can often be a short-term solution to poor weed control in a previous crop, in the longer term, problems will re-surface due to the persistence of the buried weed seeds in the soil seedbank, from which viable seeds will be brought to the surface by ploughing in subsequent years. Indeed, many weed species have seeds that are extremely long-lived due to the evolution of physical mechanisms like hard seed coats and complex survival strategies like dormancy. Consequently, if the land is ploughed to the same depth in successive years, viable seeds will be re-inverted and will germinate if conditions are suitable.

Reduced tillage, the concept of direct drilling crops without resorting to ploughing, became popular after the development of the non-residual herbicides, mainly as a method of reducing costs, but has recently been the subject of renewed interest, primarily out of concern for soil conservation and enhanced carbon sequestration, but also for other benefits. Under reduced tillage there can be better control of soil erosion, improved conservation of soil moisture, positive effects on soil biological activity and more efficient use of fossil fuel. However, wind disseminated and perennial weed species can increase, and volunteer weeds can also become a problem. In addition, not all soils are suitable for reduced tillage and less nitrogen may be made available to crops where cultivation is reduced to a minimum.

A considerable amount of research has been done on looking at the effects of tillage on weed control and in comparing the merits of ploughing with reduced tillage systems for weed management. Some of this work is summarized below as an indication of the underlying principles and as a guide to the types of effects that can be seen.

Tillage method can alter distribution of weed seeds in the soil profile. Tillage regimes that invert the soil will tend to distribute seeds through the soil profile to the ploughing depth, whilst non-inversion or minimum tillage systems will tend to leave seeds near the surface (where they can potentially be depleted by shallower cultivations). Observations show that working soil to a depth of 10cm (4in) can be normally expected to carry 20 to 40 per cent of seeds to between 5 and 10cm (2 to 4in) below the soil surface. A range of studies has shown broadly compatible results. In one, 60 per cent of weed seeds were found in the surface 1cm (0.4in) of soil in a non-tillage system. Where the soil had been chisel ploughed, only 30 per cent of seeds were in the top 1cm (0.4in) of soil (and density then dropped off). Where mouldboard ploughing had taken place, there was a uniform distribution of weed seeds in the top 19cm (7.5in) of soil. A comparison of cultivation regimes in corn (*Zea mays*) and corn-soybean (*Glycine max*) rotations over a three year period showed that the weed seeds in the top 10cm (4in) of soil following non-tillage, chisel ploughing and mouldboard ploughing represented 74 per cent, 59 per cent and 43 per cent of the total seeds, respectively. After a seven year period with the mouldboard plough and ridge till systems, 37 per cent and 33 per cent, respectively, of seeds present were found in the surface 5cm (2in) of soil. With the chisel plough and non-till systems, 61 per cent and 74 per cent, respectively, of seeds were present in the same surface 5cm (2in) layer.

Tillage method can affect weed seed reserves in the soil. The long-term decline of weed seed reserves, or weed seedbank, in soil will generally be less in the absence of cultivations, although the response of the weed flora to reduced cultivations will depend, to a large extent, on the balance between the buried seed reserves and freshly shed seed, as well as how often, and in what manner, the soil is mixed. The importance of seed–soil contact in stimulating germination is well known, and differences between cultivations may affect moisture retention and temperature fluctuations in soil, as well as altering the microtypography. This will have a differential effect on species with seeds of different shapes, sizes and seed coat characteristics, which will ultimately influence the species composition by affecting seed germination and seedling establishment. For instance, seeds with rough or mucilaginous seed coats may be better able to germinate in drier conditions; seeds with a requirement for light may also benefit from reduced cultivations. In a series of experiments over nine years using different primary cultivations in a vegetable crop rotation, seed numbers of annual meadow grass were 17, 27 and 57 million per hectare (7, 11 and 23 million per acre), respectively, for deep-ploughed 38cm (14–16in), shallow-ploughed 17cm (6–7in) and rotary cultivations 17cm (6–7in).

Tillage methods can influence the type of weeds present. By preferentially favouring some growth habits or forms of propagation, tillage can influence the types of weeds present. For instance, perennial weeds are seriously affected by ploughing but often thrive under reduced tillage. The perennial grasses, common couch and black bent (*Agrostis gigantea*), show some growth modifications under minimum tillage with the rhizomes growing closer to the soil surface or even over it, where a litter layer has built up. Similarly there is usually an increase in annual grass weeds associated with reduced cultivations and the decline of some broad-leaved annual weeds. Perennial weeds are thought generally to increase over time in organic farming systems and, depending on the weeds involved, it may be necessary to plough periodically to keep them at a manageable level. In one study of tillage for preventive weed control, plough based methods were better than harrow based methods in grain crops and, although annual weeds were not a serious problem, shallow cultivation resulted in more weeds than deeper cultivation. Infestations of the perennial grass weed, common couch, were also greater following shallow tillage.

The influence of tillage methods on specific weeds has been the focus of many studies, although they tend to neglect the wider system implications of any observations. In the UK, it has been found that annual broad-leaved weeds were less influenced by tillage than annual grass weeds. Annual meadow grass, wild oat and black grass (*Alopecurus myosuroides*) were all favoured by non-ploughing techniques. Barren brome (*Anisantha sterilis*) is said to be almost completely controlled by ploughing. In cereals, reduced cultivations do not necessarily reduce overall weed biomass but can alter the relative importance of individual weed species. For instance, the early sowing of winter cereals associated with minimal cultivations and the use of tine cultivations has favoured black grass. Similarly, certain broad-leaved weeds increase in importance including groundsel, shepherd's purse (*Capsella bursa-pastoris*), parsley piert (*Aphanes arvensis*) and the mayweeds (*Matricaria* spp.). Knotgrass (*Polygonum aviculare*) and fat hen may become the dominant species according to some researchers but others have found that fat hen and annual *Polygonum* spp. decrease in the absence of cultivation. In these cases the disparity may be accounted for by differences in loss of seeds on the soil surface due to predation and post-germination losses in the different studies. Many species show no response to changes in cultivation regimes, including common poppy (*Papaver rhoeas*), common field speedwell (*Veronica persica*) and scarlet pimpernel (*Anagallis arvensis*).

Species that increase vegetatively, like couch, will also tend to benefit in the absence of cultivations. Others include onion couch (*Arrhenatherum elatius*), creeping thistle, field bindweed and the docks, all of which can also benefit. Biennial species that are normally not adapted to cultivated land may increase if cultivation is less vigorous. Wind-borne seeds, both perennials such as sow thistle (*Sonchus arvensis*), willow herbs (*Epilobium* spp.), ragwort and dandelion, as well as annuals like wild lettuce (*Lactuca* spp.) and smooth sowthistle (*Sonchus oleraceus*), can be opportunistic colonizers of reduced cultivation systems. In reduced cultivation systems, seed shedding by crops at harvest may result in an increase in volunteer weeds. In contrast, many of the current annual weeds were introduced originally as seed contaminants and are not adapted to dispersal other than by the actions of the farmer; these species will tend to decline under reduced cultivations.

Other practices associated with reduced tillage might also have effects on the weed flora. For instance, stubbles left after harvest will attract seed- and fruit-eating birds, which will increase seed predation but will also increase seed dispersal in bird droppings. Seedlings of black nightshade (*Solanum nigrum*) and elder (*Sambucus nigra*), for example, could become more common.

Secondary Tillage

Secondary tillage is used to prepare seedbeds and leave a level surface for drilling. Typically it involves disking or harrowing to a depth of 10cm (4in). Rotovators and power harrows are also used and are able to prepare seedbeds, even when ploughing has not been carried out. Implements are available that can combine shallow seedbed preparations with some deeper cultivations in a single pass. Others can loosen the soil below the surface while leaving the preceding crop debris on the soil surface. Secondary tillage practices often present a good opportunity to control weeds, usually by manipulating the timing or technique used for maximal impact on weed populations.

Timing

Seedbed cultivation should be targeted to avoid the peak flush of weeds that may be problematic in any particular crop (except if preparing a stale seedbed, *see* below). It is well known that sowing autumn cereals as late as possible allows black grass to germinate and be controlled before the cereal crop is established. Like black grass, many other weed species emerge only at particular times of year (*see* Chapter 5). Delaying drilling until mid-October may reduce disease problems as well as weeds, but germination and growth of the crops can be slow, making them vulnerable to slug attack. Timing is also important in relation to the size of the weeds and how well established they are (*see* Chapter 3), with younger weed seedlings being easier to disturb or bury than larger weeds.

Method of Preparation

It may also be possible to tailor cultivating implements to the weed flora, since weeds are affected differently depending on the implements used. For example, spring tines will tend to drag out weeds like chick-weed but just move around better rooted weeds. Studies have also shown that the spring tines tend to move seeds upwards in the soil profile, whilst rotovators have the net effect of moving seeds further down the soil profile (*see* above). Tines are therefore better for retaining freshly shed seeds on the soil surface, allowing them to be 'flushed out', rather than incorporating them into the weed seedbank.

It is also possible to design strategies that take advantage of weed control across a rotation. For example, one novel two-year cultivation strategy has been developed in non-inversion tillage systems. In year one, the inter-rows of a cereal crop are kept weed free with repeated use of a tractor hoe that also stimulates seed germination and reduces the weed seedbank. In year two, row crops are sown where the previous inter-rows have been, while the previously cropped strips become the new inter-rows and are cultivated accordingly. In this manner, crops are sown into weed free areas from one season to the next.

Stale (or False) Seedbed

A stale seedbed is prepared several days, weeks or even months before planting or transplanting a crop. This technique combines timing and soil preparation to achieve a good first strike on weeds before the crop is established and is recognized as a strategy suitable for organic farming, which has been widely used for many years. The stale seedbed is based on the principle of flushing out weed seeds that are ready to germinate prior to the planting of the crop, depleting the seedbank in the surface layer of soil

Fig 10 A flush of weeds on a stale seedbed.

and reducing subsequent weed seedling emergence. The soil is cultivated about four weeks before drilling to stimulate germination and encourage the first, and usually biggest, flush of weeds. If irrigation is available, it could be used to help 'flush out' the weeds, otherwise the technique is very dependent on rain to stimulate weeds to germinate. Delaying sowing extends the stale seedbed effect. Once the weeds emerge they can be flamed or very gently cultivated to remove them immediately before drilling, giving the crop a clean start. In some cases the weeds can be flamed post-drilling but pre-emergence of the crop but this technique should obviously be used with care! Although adequate moisture is vital in determining the efficacy of the stale seedbed technique, soil factors such as the fineness of the seedbed and prevention of capping are also important for maximizing weed seedling emergence.

Stale seedbeds can be an effective method of decreasing the density of annual weeds, as has been demonstrated in many studies. The technique depends on weather and rotation design. A tight cropping schedule may not allow for a break of several weeks, or weather conditions may be unsuitable; for example, too cold or dry to stimulate germination. Studies have shown that when conditions were moist, 50 per cent of the weed seedlings (expressed as a percentage of the total seedling emergence in a sixteen-week period) emerged within six weeks of cultivation. In contrast, in drier years, 50 per cent emergence was related to rainfall events, sometimes as much as thirteen weeks after the initial cultivation event. If the land is prepared early, and the weather is very dry, the positive impact of early cultivation may turn negative as the land may be at risk from erosion and also lose what little moisture was available.

There are a number of points to have in mind when using the stale seedbed technique in organic systems including the following:

- Removal of emerged weeds needs to be delayed until the main flush of emergence has passed to have maximal effect.
- It may be risky to delay planting or drilling if soil conditions are good and there is a chance of heavy rain preventing future operations.
- If there is no rain during the period there can be increased soil erosion and soil drying. The resulting dry seedbed conditions and delayed crop establishment can reduce crop yield.
- Once the weeds have emerged they must be killed or removed by an acceptable method. Emerged weeds can be controlled by flaming or light cultivation or undercutting. It is important not to cultivate below the top 1–2cm (½–¾in) soil, otherwise a further flush of weeds may emerge.
- To gain the most advantage from the technique, the seedbed needs to be weed free at the time of crop planting or drilling.

Other methods may also increase the chances or effectiveness of stale seedbeds. Punch planting makes use of the stale seedbed technique but minimizes soil disturbance further (and therefore decreases the chance of additional weed germination) by dropping the seed into holes made by a dibber. This can be especially suitable for vegetables like leeks. Covering soil with polyethylene sheeting is also known to increase weed emergence and it has been looked at as a way of improving upon the stale seedbed technique. In one study, covering the soil with clear polyethylene increased weed seed germination, but germination continued after the first flush of weeds had emerged and after the covers were lifted. In contrast, following the removal of the black polyethylene, the ground was clear due to seedling death or a lack of emergence.

Cultivation in Darkness

A novel method of reducing seedling emergence, and one that has received considerable research and farming press attention, is to carry out seedbed preparations in the dark to avoid stimulating weed seed germination. It is known that light can break weed seed dormancy and stimulate germination, and cultivation in the dark has been shown to reduce weed emergence by up to 70 per cent. Covering soil with opaque material immediately after cultivation, or leaving it exposed to light, makes no difference to weed emergence, suggesting that it is the exposure to light during the cultivation operation that is important.

In practice, cultivation in darkness is usually much less effective, as it still leaves enough weeds to reduce crop yield. For example, not all seeds require light to germinate, and some may lose their light requirement with age, so the technique will not be effective on all species. Other species require very low light levels, so moonlight might be enough to trigger their emergence. The effect of light on germination will also depend on other environmental factors, such as soil moisture and fertility levels, and also soil temperature. In one study, the emergence of common chickweed and fat hen was reduced by cultivating in darkness but that of black grass was unaffected. In trials with carrots (*Daucus carota*) there was only a small difference in weed numbers between seedbeds prepared in darkness or in light, and then drilled normally. After intra-row brush weeding, there was little difference between carrot crops drilled in the dark and others drilled in the light. Daytime cultivation with a mouldboard plough has been shown to stimulate weed seedling emergence 200 per cent above that of night-time ploughing. However, when a chisel plough was used, weed emergence was much greater overall and there was no difference between night and daytime cultivations. In spite of the variable results, it seems as if dark cultivation may at least delay weed emergence. In crops where weed control is critical, the practice could widen the window of opportunity.

Following the generally disappointing results from studies in the UK, a number of potential areas of improvement in the method have been highlighted. One suggestion has been to roll the soil following cultivation, to consolidate the seedbed and prevent light penetration into the top few millimetres of soil. It is not necessary to work the soil in total darkness – covering the cultivating implement with sheeting to prevent light reaching the soil at the point of cultivation may be sufficient. The covering of tractor lights, with green filters, has also been reported to be effective. Alternatively, guidance systems may allow a range of operations to be performed in complete darkness.

Tertiary Tillage

Tertiary tillage is the soil cultivation that is used directly as a means of physical weed control, often involving a range or purpose built weeding equipment or machinery. It is dealt with in some detail in Chapter 3.

Post-Harvest Tillage

An important additional consideration when using tillage to aid weed control is the timing of any form of post-harvest soil cultivation, in relation to its effect on the movement and persistence of weed and crop seed shed during or after crop harvest. The burial of recently shed seeds can induce dormancy when conditions are not appropriate for germination; for example, the burial of winter barley (*Hordeum vulgare*) seeds in dry soil can induce dormancy and cause problems in later cropping sequences. Post-harvest cultivation too soon after seed shedding, and in sub-optimal conditions for germination, can instil a light requirement and, as a consequence, induce dormancy and persistence in oilseed rape seed shed during crop harvest. Not all seeds have the same response – barren brome seeds left on the soil surface persist longer than those buried soon after shedding. In this instance, early cultivation would be more appropriate to ensure control.

Cultivation as soon as practicable after harvest is also recommended for the control of rhizomatous grass weeds, such as common couch and black bent. An intensive rotary cultivation is needed to work the soil to the full depth of the shallow rhizome system. The aim is to fragment the rhizomes as small as possible and this works best in previously undisturbed soil. After the initial cultivation, further passes at this time only serve to move the broken rhizomes pieces around. Fragmentation stimulates regrowth of a dormant bud on each rhizome fragment. Cultivations to control regrowth may be repeated every two to three weeks or when the grass has leaves 5–10cm (2–4in) long, until no further regeneration occurs. Alternatively, the land may be deep-ploughed to bury any regrowth below the depth it will emerge from.

While stubble cleaning may not be appropriate for dealing with the shed seeds of some weed species, it can be an effective way of controlling some important weeds including charlock (*Sinapis arvensis*), common chickweed, groundsel, wild radish (*Raphanus raphanistrum*), shepherd's purse and some speedwells (*Veronica* spp.). The surface soil should be cultivated to a depth of not more than 5cm (2in), and this operation repeated at fourteen-day intervals. Some farmers take this a stage further and prepare a seedbed at this time but leave it unsown until spring. The land is cultivated at regular intervals to deal with seedling weeds and then ploughed after Christmas in preparation for spring cropping. Weeds controlled by 'autumn cleaning' include black grass and charlock. Nutrient leaching is likely to be a problem in soil left bare over-winter and the method may not be allowed under agro-environmental schemes, as there are benefits to birds of over-wintered stubbles.

Where there is a two to four week period in the autumn between the harvesting of one crop and drilling of the next, there is an opportunity for early tillage operations to stimulate weed emergence and for those weeds to be controlled mechanically before final seedbed preparation.

Success depends, though, on climatic conditions. In dry conditions, the burial of shed cereal grains can induce dormancy in them, causing volunteer weed problems in later cropping sequences.

Other Tillage Methods

A number of other tillage approaches can be useful as part of weed management strategies. Ridge tillage, for example, is widely practised in crops like potato, which can be planted in ridges. Ridges can be knocked down and reformed once or twice during the cropping cycle as a method of managing weeds and a range of machinery exists for doing this. Punch planting has already been mentioned as a method of improving the effectiveness of stale seedbeds, as it minimizes soil disturbance prior to sowing or planting, and can be used in combination with other weed control techniques, such as flame weeding. Systems of mulch tillage can also be used that rely on various methods of killing living manures to leave a smothering mulch layer on the soil.

The ultimate in reduced cultivations is no-till or zero tillage, which usually relies on the use of broad spectrum contact herbicides to deal with the weeds. Organic farmers may, therefore, be unable to practise energy saving,

Fig 11 Ridging potatoes helps with weed control.

non-inversion tillage. It has been suggested that wild oat would be less persistent under no-till, where seeds are left on the soil surface open to predation and post-germination mortality. With zero tillage there is a build-up of plant residues on the soil surface that may have benefits for soil and moisture conservation. There may be an effect on soil temperature that could delay crop germination and early growth. The residues may have an allelopathic effect and may reduce soil fertility, at least initially. The use of allelopathic cover crops in reduced tillage systems may provide additional weed control benefits. Winter rye, grown to increase soil organic matter and protect the soil, leaves a residue able to reduce weed biomass by over 60 per cent. The effect of the build-up of crop residues on pests and diseases is not known, but there are thought to be benefits.

LIVESTOCK AND MANURES

In mixed systems, where grass–clover leys are used for fertility building and as pasture, livestock can make good use of the nutrients and produce a valuable resource, which can be used around the farm to fertilize cash crops. Pastures will also need weed management. Conservation of old or rough pastures is becoming increasingly widespread and may require specialist grazing. Animals can also be used to consume cut weeds or other plant material like chaff or screenings that are likely to contain some weed seeds.

Livestock

Animals have different grazing behaviour and it is even recognized that different breeds or individuals are likely to have different tastes and habits. The species, breed, age and individuality of animals will all affect what they will eat and therefore what effect they will have on both weeds and pasture. Variability within the feeding site (for example, vegetation, topography) can also be important, as can other factors such as the weather. Although it is impossible to detail all the grazing characteristics of farm animals, in general terms the following applies:

- Sheep are recognized as being useful for weed control as they graze close to the ground and will eat a wide range of plants. They can be used early, and some breeds are hardy and will graze well under a wide range of conditions.
- Goats are browsers and have a reputation for enjoying tough and woody plants but are not commonly used except in specialist production systems.

- Cattle can be used for early grazing but there are a large number of dif ferent breeds and types with different grazing requirements, includin beef, dairy and traditional breeds. Grazing strategies appear to be relate to plant energy content and digestibility and this will affect how plant are eaten (leaves or stems or other parts of plants), which plants ar eaten (species) and size of plants eaten (young or more mature plants) Cattle tend to avoid longer, coarser grass and hairy, spiny or poisonou plants. The selection of certain plant species and plant components, a well as the location of these plants, is based on the previous experienc of the animals or learned from their mothers when they are calves.
- Pigs are good at rooting and have been recommended at various time for digging out perennial weeds, like dock and couch, when fence within fields (and tightly stocked). They are also useful for rooting ou volunteer potatoes. They can however have a negative impact on soi structure, especially if kept *in situ* for a long period in wet soil condition
- Horses and ponies are grazed on ever increasing areas of land and ca be used as part of a grazing rotation. They prefer frequent small amount of fibrous grass or other high-roughage material. They can be pick eaters and selectively leave many species and have, for example, bee known to dramatically increase the number of docks in a rotation.

Fig 12 Pigs can be used to root out weeds, but may damage soil structure.

- Geese consume grassy weeds and have even been used to weed in between rows of well-established crops.

Weed management includes getting the right grazing balance over the year and, ideally, alternating grazing species from year to year. Grazing regimes will necessarily be tailored to fit individual farms and circumstances, but suggestions for rotating livestock, depending on situation, could include using sheep to graze out certain weeds such as ragwort, or free-range outdoor pigs to forage for roots, which could reduce the perennial weed burden. Sheep may also be useful to lightly graze oats to encourage tillering and hence weed suppression potential. It is important to get the right grazing balance over the year to get the maximum benefit for the animals and also to prevent damage to the sward or soil; for example, stocking more lightly in the winter months and in wet periods prevents poaching.

In summary, there are a wide range of important factors that need to be assessed when using animals to graze or manage weeds. Animals will need to be grazed in a suitable season, the length of grazing period should be appropriate, as should the stocking density. Finally, a recovery period for the grazed area should be factored into any grazing regime. Things to consider include:

- timing grazing to benefit the pasture and promote competition with weeds;
- timing grazing to damage the weeds (for example, to remove seedlings, flowers or seed heads before seed production);
- allowing time for the pasture or forage to recover between grazings;
- making sure that livestock that have been grazing on weedy land feed on weed seed free forage for four to five days before introducing them to weed free areas or pastures (some seeds will remain viable after passing through animals which may take a few days).

Use of Manures

The use of raw manures and slurry has often been associated by farmers with increased weed problems. The problems can arise in various ways: as a result of weed seeds in the manure or slurry; as a result of the way in which it is applied; or due to the stimulatory effect of the nutrients on weeds already present in the soil.

Manure may contain weed seed, either seed that has passed undigested through animals or from bedding materials like small grain straw and old hay. High temperature aerobic composting (recommended under organic

Fig 13 High-temperature windrow composting will kill weed seeds.

standards) can greatly reduce the number of viable weed seeds, as long as the temperature is maintained at higher than 60°C (140°F) for more than three days. Operationally, compost will need to be regularly turned to achieve even heating through the whole heap and to get material from the outside (where seeds are likely to survive) to the inside (where the highest temperatures are likely to be generated). In a similar way, aeration of slurry can reduce the number and viability of weed seeds.

Applying manure can be an opportunity for weeds to establish. It is advisable when applying manure or slurry to try not to create conditions that stimulate weed seeds to germinate, for example, through excessive soil disturbance, creating bare patches. Applying slurry to stubble after silage cuts can provide optimum conditions for weed seed germination. A nutrient rich bed of cattle slurry will produce a high potassium environment, which will favour weeds such as docks rather than grasses. Dock seeds should not survive low pH silage but will survive in a later cut of hay.

Some research has shown that placing manures and slurry more accurately on crops can benefit the crop rather than the weeds. Crop plants are generally sown fairly deeply and they germinate from a lower level in the soil profile than weeds, which tend to dwell on the surface and germinate from 0 to 3cm (0–1in). Crop plants also root more deeply. This tendency can be exploited for weed management. In arable/horticultural systems, manure placed 10cm (4in) below the soil surface encourages the crop seeds to grow down into the nutrient rich layer before the surface dwelling weeds can reach it. This technique can also be used with broadcast spread slurry. If it is ploughed in rather than left on the surface it will be available to the crop before the weeds can reach it.

In many cases, the growth of weeds that follows manuring is a result of the stimulating effect manure has on weed seeds already present in the soil. This can be due to the flush of nutrients, such as the supply of nitrates, enhanced biological activity in the soil or other changes in the fertility status of the soil. Some work has indicated that excesses of potash and nitrogen, in particular, can encourage weeds. In any case it is prudent to monitor the nutrient content of the soil and manure, and spread manure evenly to reduce the incidence of weed problems. In this respect, composting can enable more even spreading, as the final product is more crumbly.

FARM HYGIENE

Farm hygiene is an important part of the whole farming system but it is often overlooked when planning weed management programmes. As part of a weed management strategy, hygiene aims to prevent weeds spreading between fields on the farm and establishing in areas where they are not already present. Regardless of how well weeds are controlled within the field, fresh weed seeds may still enter from external sources and provide additional weed problems, and it may be possible to manage this to some extent by ensuring farm imports are free from weed seed. Some of these factors are considered in this section.

Early Detection System

Weeds are, by their nature, tenacious and difficult to eradicate once a population becomes established. The best form of management is preventing the establishment in the first place; this will be greatly aided if a particular weed can be detected early and prevented from spreading. To this end it is worth ensuring that all people who work on the farm, or visit it, are encouraged to report areas of problem weeds or any new weeds that have appeared. Following on from this it is vital to keep records of weed problems and their extent in order to be able to monitor any spread or otherwise. Simple field maps (often downloadable from the internet) can help to monitor the change in weed populations, if updated regularly. It is not necessary to record detailed numbers but rather use a scoring system or take notes that can speed up the process. A digital camera can also be a useful in this respect as a (relatively pain free) tool to record presence of weeds and monitor changes over time, especially if pictures are taken from fixed points in the field at different seasons and years.

Farm Equipment and Machinery

Field staff and visitors should be alerted to the possibility of spreading weeds and weed seed. Although it may seem insignificant to worry about transferring a few seeds from one location to another, the enormous capacity of many weed species for producing seeds or other viable parts can rapidly establish a population from even one or a few plants. Weed seeds can be spread on clothing and boots, and when moving between fields, shaking down clothes and scraping boots can help minimize such transfers.

If there is a serious weed infestation in a particular field, or machinery is moving through fields with flowering weeds, then washing down machinery should be a serious consideration. This may be particularly important at harvest where some weeds may be setting seed in fields (*see* below). Contractors are increasingly used for on-farm operations and their vehicles are a potential source of weed seed as they are transferred between holdings. In fact farm vehicles have been repeatedly shown to harbour large collections of weed seeds and it is worth cleaning them down regularly.

Crop Seed

In the past, contaminated crop seed was the major source of introduced weed seeds and it continues to be an important agency for the spread of weeds. There have been considerable developments in seed cleaning and accompanying legislation that have reduced the presence of weed seeds in crop seed. The decline of several formerly common weed species, such as corncockle (*Agrostemma githago*), can be directly attributed to improvements in the seed cleaning process. The UK 1922 Seed Regulations prohibited the sale or sowing of cereal seed containing more than 5 per cent by weight of five specified injurious weeds. The 1961 Seed Regulations introduced a requirement for the declaration of the percentage of all weed seeds present in seed that is sold. However, some weed seeds still get through and, if sown with the crop, the resulting weed seedlings will emerge in the crop row. Where seed is produced outside the UK, it has provided a route for the introduction of alien species or of common weeds from a different genetic background. If the alien weeds find the local conditions favourable, they may multiply to become a future weed problem. Repeated introduction can ensure the survival of a species that is at its geographical limit.

It has become a requirement of organic farming standards that all crop seed is grown organically. Based on the assumption that organic farmers tolerate, rather than eradicate, weeds, organically grown seed crops will have even greater potential for weed seed contamination than their conventionally grown counterparts. There are significant attractions for cereal farmers in using home-saved seed, including cost savings, availability and adaptation to local conditions. In 1971, farmers' own seed represented about a quarter of the conventional cereal seed sown in the UK. In a survey of cereal seed in drills on farm in 1970, 89 per cent of farmers' own seed was contaminated with weed seeds compared with 36 per cent of merchant seed. Weed seed numbers were much higher in the farmers' own seed.

Crop harvest is a critical time for the dispersal of crop and weed propagules (seed or other viable plant parts). In cereals, it has been estimated that, on average, 40 per cent of weed seeds have been shed by the time of harvest. About 5 per cent of seeds remain at below normal stubble height, leaving between 45 and 70 per cent of weed seeds to pass through the combine harvester. The combine can aid both the re-introduction and spread of crop and weed seeds to other parts of a farm. Weeds maturing at the time of crop harvest, and at a height intercepted by the combine, will have a proportion of their seeds re-introduced into the field. Other seeds may remain lodged on the combine to be deposited at a later time and possibly at great distance from the parent plant. The magnitude and distribution of these seeds is dependent on the type of combine. Small seeds, in particular, are redistributed in the field. In the UK, modification of combine harvesters to separate out weed seeds from grain and straw, to avoid returning seeds to soil, has been recommended. Unfortunately the collection of seeds and chaff slows down harvesting, increasing costs and is not always practised.

Crop seeds lost during harvesting can also be dispersed to emerge as volunteer weeds in subsequent years. Seed shed during the harvesting of oilseed rape (*Brassica napus*) can give rise to over 500 seedlings per square metre (per 11 square feet) in following crops. Crops that reproduce vegetatively can also leave behind propagules after harvest; for example, volunteer potatoes result from the small daughter tubers that escape the harvesting process. Volunteer weed seeds can be a particular problem in harvested plant material. Cereal straw used for mulching often contains shed grain and sometimes whole ears of wheat or barley. If the straw is from a weedy crop, weed seeds may also be present. Identification and elimination of modes of re-introduction and spread of weeds through the harvesting process, offers a substantial area of improvement for reducing potential future weed populations and should form part of any weed management strategy on an organic farm.

Fertility Inputs

Organic materials that are composted on and off farm, such as green waste, animal bedding and manure, can contain large numbers of crop and weed seeds before treatment. It is essential that any compost, manure or slurry that is to be spread on fields has been processed correctly and is free from viable weed seeds. If compost is made on the farm, the heap will need to be turned regularly to ensure the temperature rises high enough throughout the material to kill any weed seeds present. The temperature needs to remain at 60°C for three days to ensure that the heating process has been effective. Ideally, after treatment the compost or manure should be covered to prevent contamination with wind borne seeds. Heaps stood out in the field should be kept weed free to make certain no weed seeds are shed on them. Slurry should be adequately aerated to ensure weed seeds have been killed. Manure and slurry are best applied to grass in spring and summer, often following a conservation cut. Slurry application may cause a temporary check to grass growth that can give weeds an advantage. In addition, cattle slurry is high in potassium, which is thought to encourage some weed growth, for example in docks.

Livestock Feeding/Watering

Water and feeding points must be rotated to ensure that excessive dunging does not occur. Perennial weeds like thistle and docks thrive in high-nutrient areas. If livestock have been grazing on these plants, viable seeds can pass through their digestive systems and germinate when conditions are favourable. Over-grazing and pinchpoints (for example, around gateways) can also be problematic areas, and care should be taken to try and reduce poaching around these areas to a minimum to prevent weeds establishing (*see* above).

Management of Non-Cropped Areas

Weeds on wasteland, or old manure heaps around the farm, should be prevented from seeding. In some cases this may have to be balanced with biodiversity needs, such as encouraging beneficial insects on particular flowering weeds. Management of open-water sources can also have implications for weed management. If dirty water is spread on the farm, it

Fig 14 Reinforced area around watering point.

Fig 15 Field margins may need careful management to prevent weeds encroaching.

should be from a covered source or seeds from weeds should be prevented from entering it. Many farmers feel that spreading water contaminated with weed seeds has been an important source of weed spread around the farm. This might particularly apply when water from ponds is used for irrigation.

Areas of the farm, which are not cropped, still need to be managed to prevent the spread or invasion of weeds. Many farmers rigorously top or remove weeds like dock from hedgerows and fence lines to prevent seed moving into the field, but any such measures should be checked to ensure that they conform to any environmental schemes that might apply. Field margins may provide a valuable corridor for movement of flora and fauna and it is an area of importance to a number of threatened plant species. They can also be reservoirs of the natural enemies of crop pests and there is data on the movement of beneficial arthropods from the margins into the crop. It has been shown that some perennial weeds from field margins are less invasive if a strip of grass or wildflowers is sown around the field boundary.

Chapter 3

Direct Weed Control

Although the cultural methods described in the previous chapter provide the underpinning for organic weed management, it is likely that direct action will, at one time or another, be needed against weeds to prevent crop losses. By direct action we mean that weeds will have to be physically killed in, or removed from, the crop or field. Before taking direct action, the weed flora should be viewed in the context of the damage it is likely to cause in the current crop (*see* Chapter 5), the long term implications of not taking any action (*see* Chapters 2 and 5) and the likely costs and benefits of any immediate control measures (*see* Chapter 4). For instance, a low weed population may have little effect on the current crop but could shed massive numbers of seeds, adding to the weed seedbank and exacerbating future weed problems. Farmers and growers will therefore need to consider whether weed control is justified and, if so, when and where that control is needed. Having established that direct action needs to be taken, there is the question of which method of control should be used and of deciding how it fits into the overall weed management strategy for the farm. The aim of this chapter is to discuss these aspects of weed control on organic farms.

IS DIRECT WEED CONTROL NEEDED?

A combination of factors will influence the decision on whether or not to take direct weed control measures. A good working knowledge of the crop, the potential weed flora and, ideally, previous weed management experience on the site will be indispensable as a guide to any action to be taken. An understanding of the biology of the particular weeds will help in deciding how to deal with them appropriately (*see* Chapter 5). For example, it is particularly important to know whether the main weeds are annual, biennial or perennial, as this will influence the method of weed management chosen. In the latter case, it will be particularly important to know what specie(s) are present, as perennial weeds have many different survival strategies and successful control will need to target any weaknesses in the life-cycle of any particular weed.

There are no hard and fast rules when deciding to take direct action against a developing weed problem. Perhaps the most important things to consider when deciding if weeds are a problem and if control is needed are:

Will marketable yield be affected by the weeds? Yield reductions occur when weeds compete directly with the crop and, without some form of weed control, marketable yield can be severely reduced. Direct yield losses can be important in any crop and will negatively affect the profitability of any crop enterprise; for instance, reductions in fresh weight of as much as 96 per cent have been reported in weed trials with salad onions. Vegetable markets, in particular, often have exacting growth and size requirements, which may seriously be compromised by weed competition over and above any direct yield loss.

Are the weeds having other negative effects? The physical presence of weeds may impair the harvesting process or affect the marketability of the crop. For example, cleavers can directly impede harvesting equipment in cereals. Weed parts or seeds can contaminate produce. Shed seeds may affect the quality of the crop if, for example, seeds fall into lettuce heads or contaminate crops such as cereals. Weeds may also encourage crop pests or pathogens, but the direct effects will be more difficult to judge.

What are the long-term consequences of infestation? Lack of control may result in a build-up of the weed seedbank or spread of perennial weeds, even if there are no short-term direct or indirect reductions in yield. In this case, immediate weed control may be justified to reduce future costs. Once again, it will be hard to judge outcomes or consequences in many cases and the best guide may in fact be past experience with particular weed species or fields.

Are there any benefits associated with the weeds? Often the intangible positive benefits of weeds are the hardest to judge but this should not let these real benefits be ignored. Some weeds in grassland may be a useful source of trace elements, for self-medication or for extending the grazing season, although many suppress grass yield by a significant percentage. Some weeds can deter pests like aphids or root flies to some extent, although once again there will be some competition effect on yield. The presence of 'weeds' might also be counted as a positive benefit within environmental schemes or within conservation areas in which case farmers can be expected to be compensated for any yields foregone.

In field crops, studies of crop losses due to weed competition have repeatedly demonstrated that the cost of weed control can outweigh the value of any yield benefit. The notion that the cost of control should equal the financial benefit to the crop has led to the concept of economic thresholds for weeds, whereby weed species are classified according to their relative

Fig 16 Is weed control needed in this lettuce crop?

competitiveness in a particular crop. For example, a cereal may be able to tolerate twenty field pansies (*Viola arvensis*) per square metre without yield penalty, but only one fat hen plant. The cost of any likely control method can then be calculated and compared with the value of any potential yield loss. However, such thresholds are generally found to have some serious shortcomings; they do not generally take into account the combined effects of different species present at different densities in the weed flora, nor do they take into account the potential build-up of the weed population if control measures are omitted and the implication this may have for future crops. They can also be inflexible in their predictions in field situations; for example, if the weather conditions change. This has limited their use in practice and led to weed management strategies that err on the side of caution; that is, it is better to proactively control weeds when the opportunity arises, rather than waiting for supposed thresholds to be reached.

Also, whilst it might be relatively easy to give answers to these questions in arable and horticultural systems, in grassland the decision to take direct action against weeds is more difficult to make. Apart from poisonous plants, like common ragwort, the distinction between crop and weed is often unclear. The so-called 'weed grasses' may contribute to the sward by extending the grazing season. Other weeds may be a useful source of trace elements, may be sought out by animals for self-medication or make a valuable contribution to biodiversity. Although broad-leaved weeds suppress grass yield to a significant degree, they often add to the overall yield when the crop is cut for silage.

Two tips emerging from the knowledge that has been collected on organic weed management are:

- Remember that removing all weeds is not only difficult, it is not always desirable and it does not necessarily make sense economically. Many pieces of research have shown that the cost of removing weeds completely can financially outweigh the resulting yield benefits.
- Thinking long-term is crucial in organic farm systems. Planning for the long-term, while maintaining some flexibility in the short-term, is the key to successful weed management. Weed management should be considered throughout the rotation and as a property of the whole farm system. Weeds may not threaten the current crop, but may have an impact on later crops if allowed to set seed or propagate and, therefore, they should be controlled at those times that the best opportunities present themselves. The weeding of each crop in the rotation needs to be planned and the schedule of weeding operations in different crops in one season needs to be co-ordinated.

WHEN IS CONTROL NEEDED?

Once the decision to weed has been taken, the next question is when should it be done. If weeds are left uncontrolled for too long, then they may start to compete with the crop and larger weeds are usually more difficult to kill. Weeds allowed to grow on will produce seeds or other propagating parts. However, if weeding is carried out prematurely, the main flush of weeds may not have emerged and the weeding operation will need to be repeated. Weeds that emerge early with the crops are likely to be more competitive than weeds that emerge late, which tend to exert little competitive effect. However, later weeds can have a negative impact on harvesting operations and may return seeds to the soil seedbank. The timing and number of weeding operations within a particular crop are to some extent flip sides of the same coin and both can be strongly affected by external conditions, especially the weather.

Timing of Weeding

If weeds are left uncontrolled for too long, then they may start to compete with the crop. Weeding should be timed to prevent weeds directly competing with crops in the period when they are at their most sensitive, but this can depend on many factors, including the type of crop and the weed(s) present, as well as other elements such as soil conditions, weather conditions and time of the year. Generally, the earlier the emergence of the weeds as compared with the crop, the more competitive the weeds are likely to be. Weeds that emerge later in the crop cycle tend to have little competitive effect, although they can potentially return a large quantity of seeds to the weed seedbank in the soil and can even impede harvesting operations.

Number of Weeding Operations

Obviously the fewer the number of weeding operations, the lower the immediate weeding costs will be. If weeding is too early, the weeds will not all have germinated and it may be necessary to weed again later in the season to prevent weeds developing and impeding or contaminating harvest. In contrast, if weeding is carried out too late, although the number of weeding operations may be reduced, competition with the crop may have already caused damage. Weather conditions can also have a strong impact on the rates at which weeds (and crop) grow and develop, and this in turn may well influence not only the timing of weeding operations but also the likelihood of the necessity to repeat them.

Optimum Weeding Times

The optimum weeding times for different crops are not fixed and vary, not only with the crop but also the weed population and the relative times of emergence. Numerous experiments have demonstrated that, for field vegetables, a carefully timed, short, weed-free period – sometimes called the 'critical period' – or a single weeding, may be all that is needed to prevent yield loss. Early weeds are removed just before they would begin to compete with the crop, and weeds that emerge after the single weeding or weed free period do not compete with the crop (if the timing is correct). In a weak competitor like bulb onions, it has been shown that there is no adverse effect on yield from weeds present for up to five weeks after crop emergence but, if the weeds remain longer than this, they can reduce yield by 4 per cent for every day they are left uncontrolled. However, from seven weeks after crop emergence, any new weeds that emerge can be left to grow without affecting yield. This two-week period, when the crop needs to be weed free, is the critical weed free period for this particular cropping situation. In a relatively competitive crop such as drilled swede (*Brassica napus* var *napobrassica*), a single weeding operation at approximately six weeks after crop sowing may be all that is needed to maintain yields equivalent to that of a crop kept weed free throughout. Similarly, in potatoes or transplanted brassicas, which are very strong competitors against weeds, a single weeding any time within a 'weeding window' of two to three weeks can prevent losses due to weeds. The same concept has been applied in spring cereals but, in winter cereals and other crops that experience an autumn and a spring flush of weed emergence, it is more difficult to determine the optimum weeding period.

The timing of weed removal is strongly dependent on the prevailing environmental conditions, especially the weather. For instance, wet conditions can reduce the effectiveness of weeding operations and allow weeds to re-establish or stimulate new weeds to germinate. In contrast, drier conditions will prevent this. There are also some factors that can be manipulated by the farmer to give the crop a competitive advantage; for example, using module raised transplants that give the crop a head start, or by putting the weeds at a disadvantage with cultural measures such as stale seedbeds. These practices tend to widen the 'weeding window' (a period during which weeds need only be removed once) and this makes the absolute timing of weed removal less critical. Widening the weeding window in this way reduces pressure to carry out a weeding operation within a very narrow time period, which may be difficult or impossible because of weather or lack of resources – especially time!

WHERE IS CONTROL NEEDED?

If direct weed control is required, it may not be necessary, or even possible, to weed the whole field area. Weeding operations are often targeted at four specific areas within the field or crop, although any particular operation may only reach one or two of these target zones, which are:

- broad-spectrum – weeding across the entire cropped area;
- patch weeding – specific areas targeted by hand or machine;
- inter-row – weeding between the crop rows;
- intra-row – weeding in the crop row itself.

Once again, crop inspections to observe where the weeds are developing combined with previous experience, are likely to indicate where control is best focused. The action required may change as the season progresses, and as the crop and weed flora develop. Many different tactics can be used in a weed management programme and, for some crops a broad-spectrum weeding approach may be most effective, particularly the more narrowly drilled cereals, where harrows can be effectively used across the whole soil surface. In some situations it may be possible to focus on patches of weeds, if there is a particularly problematic local infestation (for example, couch or docks), and selectively remove or top them. Inter-row weeding is focused between crop rows and an increasingly elaborate array of machinery has been designed for this purpose, some of which also throws soil back into the crop row to bury intra-row weeds. Intra-row weeding is common in field vegetables but is becoming increasingly popular in drilled cereals. The increasing sophistication of machinery has also been used to develop intra-row weeders (for example, finger weeders) that remove weeds in the row with minimal damage to crop plants. Weeds within and between rows can, of course, be removed manually with high precision but costs are likely to be high.

Weeding within a crop can follow a sequence of targets. The first weeding operation is often carried out across the entire cropped area. At this early stage, crop establishment will benefit from a weed free environment to minimize subsequent competition. As the season progresses, broad spectrum

Fig 17 Weeds left in the row after mechanical weeding in onions.

weeding may continue to be effective for some crops, particularly narrow, row crops like cereals. In contrast, in widely spaced row crops, inter-row weeding may be the best option. A vigorous crop that is able to rapidly form a continuous leaf canopy will not need subsequent intra-row weeding. A more open crop may benefit from a combination of inter- and intra-row weed control. In situations where local infestations of weeds are the problem, it may be beneficial just to focus on the weed patches. Different crop and weed situations require a range of techniques, and the various options are discussed more fully in the next section.

WHICH METHOD OF CONTROL?

The method of control chosen will obviously depend upon the answer to many of the previous questions, and especially upon where the control is most needed. There are two main categories of direct control: physical and biological. In commercial organic farming, a wide range of direct physical weed control methods have been developed to compliment indirect cultural control measures (*see* Chapter 2). They are generally perceived as the most practical techniques in current use, although biological control systems might gain some importance in the future.

The key problem to be addressed is in choosing a method that removes the weeds selectively without injuring the crop. The choice of weeding method and of implement depends in part on practical aspects, such as the crop and the soil type, but economic elements like purchase price, operating costs and labour requirements are often the overriding factors (*see* Chapter 4). On small areas, or where sufficient work force is available, hand weeding remains a possibility, particularly in high value crops, but on most farms, crops are grown on too large a scale, and labour is too expensive (and often of limited availability).

Direct Physical Control

Physical weed control measures can be broadly divided into manual, mechanical, thermal and mulching. A combination of these may be more effective than relying on a single strategy. Success also depends on matching the weeding implement with the crop, the weed flora and stage to be managed.

Manual Weeding

Hand weeding is the traditional means of removing weeds from crops, either by directly pulling (hand rogueing) them or by hoeing them.

Manual methods of weed control are widely used in organic horticultural crops where it is important to perform an effective first weed to prevent early competition. Often in these crops hand weeding will follow a mechanical inter-row weeding operation with the aim of removing weeds left in the crop row.

There are a wide range of manual hoe designs, which normally have some form of cutting edge on a blade. They are used to cut weeds and stir the soil in order to dislodge or bury weeds. Many hoes are long-handled or otherwise ergonomically designed so as to minimize discomfort to the person hoeing, and so that it is not necessary to squat or kneel. However, this might be necessary in close intra-row weeding work. Perennial weeds will often need to be directly dug and pulled, in which case knee pads may improve efficiency and reduce strain.

Different types of hoe work in different ways: cutting at different depths, cutting when pulled (draw hoes), cutting on the push (push hoes) and some on both strokes (oscillating or stirrup hoes). Hoes require regular

Fig 18 Manually hoeing leeks after mechanical inter-row cultivation to remove weeds within the row.

sharpening to maintain efficiency and, in that vein, pushing is generally reckoned to require more effort than pulling. Dutch hoes are often thought of as the traditional (push) hoe but there is now a wide range of designs available. The wheeled push hoe can help to ease the hoeing process, though it is not as flexible for weeding within the row and is best set to straddle or run between crop rows. Choice can often come down to personal preference and an easy place to gain an appreciation of the choice available is the internet, although there is no substitute for actual experience.

Hand rogueing is a widely used technique for dealing with the removal of difficult-to-control species, such as docks, thistles and common ragwort in grassland. There will also be times when hand rogueing of the odd plant or patch of a particular weed is the most effective way of preventing that weed from proliferating or spreading. Hand held tools have been developed specifically to remove problem weeds like docks or ragwort (normally with prongs or forks) and weeds in difficult locations (bill-hooks). Strimmers, mechanically driven hand held devices that chop and macerate patches of weeds, have also become popular for weed control, especially in non-cropped areas, although their action is rather indiscriminate. Once again, the variety of tools available can be researched on the internet and their use will often come down to personal choice.

Manual weeding can seem a daunting task on large field areas; gangs of weeders are likely to be more motivated and work more efficiently. Gangs can walk through the crops hand weeding or hoeing but it may be more efficient to have a team of workers lying on a flat bed weeder. These weeding platforms, of typically four to eight stretchers pulled by a tractor, allow workers to lie prone to the ground while being pulled along slowly enough to weed within the crop rows. The rogueing of patches of the weeds is best achieved with gangs of four or more people moving methodically through a field, pulling the weeds and then carting particular weeds off the land.

Mechanical Control

Mechanical control methods are based largely on killing weeds by burying, cutting or uprooting them. In this they are an extension of manual methods but normally understood as involving some form of machine-mounted or machine-driven implement. Many of the machines might also serve for other tillage operations and the use of such machines for weed control is often termed tertiary tillage (*see* Chapter 2). The range of mechanical weeding implements is discussed in more detail below. With most mechanical weeding implements, operator skill, experience and knowledge are critical to success. Drawbacks to mechanical weed control

Fig 19 Tractor (front) mounted bed weeding platform to aid hand weeding and rogueing.

can, under some circumstances, include dependence on weather conditions for efficacy, delays due to wet conditions with the subsequent risk of weed control failure, as weeds become too large to uproot easily, and low work rates (as compared to other farm operations but not manual weeding!).

The weather and soil conditions at, and after, weeding have a major influence on efficacy. Moist or wet conditions will always have the potential to simply transplant weeds and allow them to continue growing. In addition, some implements are only effective on small weeds and timing is often critical. The best time to kill weeds is considered to be when they are in the white thread stage and it is sometimes said that if the weeds have emerged you are already too late with the first weeding. However, it must be borne in mind that weeds emerge over a period of two to six weeks or longer, and too early a weeding runs the risk of missing later germinating weeds so that further control passes may be needed.

BROAD SPECTRUM WEEDERS

These are weeders that control weeds across the whole field area. As such they tend to be indiscriminate in their action but can be used in a selective manner.

Harrowing is a traditional form of mechanical weeding for controlling annual weeds but is ineffective against perennial and established deep-rooted weeds. Traditional harrows are rigid in construction with steel spikes but many models now are flexible and can have a range of flexible tines (see below). They stir the soil to a depth of 2–4 cm (1–2 in) and are most effective on weeds at early growth stages, that is, up to 2 cm (1 in) in height. Weed control is predominantly due to burial of the weeds but research has shown that increasing the working depth of from 1 to 3 cm (0.5–1.2 in) doubled the number of uprooted weeds, and was further improved by higher soil moisture and faster working speeds. Dry weather is critical to the success of early harrowing operations but adequate soil moisture is needed initially to encourage rapid weed emergence.

Harrowing can be performed 'blind', or pre-emergent, before the crop has emerged, and then at desired intervals in the season. In cereals, blind harrowing is often carried out after drilling but before crop emergence in order to kill the first flush of emerging weeds with the aim of giving the crop an early advantage over the weeds and aiding selectivity in subsequent harrowing operations. Blind harrowing has little effect if few weeds have emerged, and may sometimes delay crop emergence. Harrowing is now being used more extensively in row crops, particularly in transplanted or strongly rooted crops that can withstand the uprooting action of the tines. This ensures an even working of the bed and it can be very effective if used frequently when the weeds are at the white thread (just germinated below the soil surface) or cotyledon stage.

Tine weeders are fitted with flexible tines. They work in a similar way to chain or drag harrows but the tines are formed from a coiled loop of metal or are spring-mounted. Their shape can vary in profile and flat or round tines are available. The tines vibrate through the soil and glide around obstructions. The vibrating action uproots small weeds. Flexi-tines can be used selectively at the late tillering stage of cereals, when the dense crop foliage forces the tines into the inter-row. The choice of tines depends on soil type and structure, but adjustment of the implement, especially the angle of penetration of the tines, is important. The tines are located on flexible mountings, and the angle of teeth or tines is adjustable, depending upon the degree of attack or aggression required. Tine weeders are more successful on lighter soils and are less suitable for heavy land.

Mowers, cutters and strimmers are other machines that will operate across the entire field surface. Where weeds are much taller than the crop, it may be possible to 'top' the weed and at least prevent further seeding. A machine based on a rape swather has been used as an alternative to hand rogueing of wild oats in cereals. The cutter bar is set just above crop height and, after cutting, the weed is pushed into a collecting tray for

Fig 20   A small flexi-tined harrow.

Fig 21   Flexible tined-weeder.

disposal. The machine has the potential to deal with tall weeds in other crops too. Similarly, a rotary cutter has been developed to remove the flower-heads of bolted weed beet growing in sugar beet crops. Flail and rotary mowers are commonly used to control perennial broad-leaved weeds in grassland systems. Frequency and timing of topping is important; for example, creeping thistle will regenerate more aggressively if cut in early spring (ideally this weed should be left until flower buds are just showing purple, when its reserves are at their lowest). It is felt that with vigilant topping (three to four times per year) perennial weed patches can be gradually reduced in size and number.

Other broad spectrum weeders include the 'eco-puller' which has been developed to mechanically remove perennial weeds, such as common ragwort, from grassland. The eco-puller uses a combination of belts and rollers to grip the plants and give a firm vertical pull that extracts the plants, complete with roots, straight out of the ground. The uprooted plants are then transferred by belt directly into a collection hopper for disposal. It is said to be suitable for a variety of perennial and biennial weeds on grazed grasslands, provided they are over 25 cm (10 in) high.

INTER-ROW WEEDERS

As the name suggests, inter-row implements control weeds between the crop rows. A range of machines is available, but accurate adjustment and ease of operation can be as important as the choice of equipment. Some implements are powered, others are ground driven, some are front-mounted, whilst others are rear-mounted or carried on specific tool-bars. Of the rear-mounted machines, some may require a second operator to steer as close to the crop rows as possible, whilst others have computer controlled vision-guidance systems that adjust the steering automatically. These types of implements are known as steerage hoes. A short resume of the main types of implements is provided below.

Non-powered sweeps/shares/ducksfoot hoes are usually mounted on more or less complex tool-bars that are in turn mounted on the front or rear of tractors. The main control actions of these blades are cutting and burial. There are different types of blades available and various mounting methods. Hoe blades are mainly A- or L-shaped when viewed in plan. The working depth is typically 2–4 cm (1–2 in). Normally they can be adjusted on the tool-bar to accommodate the row spacing in the crop and the working depth required. Some of the tool-bars or machine frames can be 'steered' by a second operator (so-called steerage hoes), which can improve the precision of their action. In a further step there are now a range of machines that use computer-vision systems to identify crop rows and automatically adjust the position of the tool-bar and blades in real time (*see* below).

Fig 22 A simple scuffle with rigid sweeps.

A good seedbed and precise drilling of the crop are prerequisites for successful inter-row hoeing in order to avoid excessive crop damage. For instance, the technique of harrowing-in cereal seed after drilling may displace the seed out of the row leading to crop damage during hoeing, and seed rates should be increased to compensate for any likely losses. Increasing the working depth has been shown to do little to improve weed kill, but higher forward speed increases soil covering of weeds and reduces survival. Hoes can be particularly effective against mature weeds. In order to protect the above-ground parts of plants from mechanical damage, and from being covered with soil, different types of protectors can be fitted. These may take the form of discs, plates or protective hoods. In practice, most farmers and growers will take time to adapt and set up a weeding implement that works in their farm system and which can be rapidly brought into use once conditions are favourable for weeding. This usually involves working to standard row (and bed) widths and may involve a certain amount of modification of the original specifications. A good tool-bar will be indispensable.

In order to try and improve the efficiency of weeding operations, and overcome some of the limitations, precision-guidance weeding systems are now being developed, which allow machinery to be driven closer to the crop rows and at a higher speed. These involve some kind of

Fig 23 Rear-mounted steerage hoe with rigid hoes.

computer-actuated guidance system, which can differentiate between crop and weed. Several models, which have been attached to hoes, are now available in the UK. The work rate of a 4m (13ft) wide automated guidance hoe can increase more than five-fold, and it is now also feasible to use hoe widths of 8m (26ft), or even 12m (40ft), that can increase the work rate further. The savings in operational costs have to be balanced against the higher outlay costs. With most of these systems, the limitation is still that the intra-row area is not weeded, but machines are being developed that can move into the crop rows and selectively remove the weeds.

Non-powered rotary cultivators are normally driven by the forward momentum of the tractor and include machines like the basket or cage weeder. This machine has two horizontal axes on which the rotating baskets are mounted. These axes are connected via a chain and sprocket arrangement causing rotation at slightly different speeds, the first driving the second. Telescoping baskets allow adjustment of the row spacing. The baskets uproot weeds in the upper 2.5 cm (1 in) of soil or strip leaves from weeds. They are particularly effective when used in conjunction with an inter-row hoe that breaks the soil surface beforehand. They are effective where there are big stones or long stemmed residues.

Fig 24 An inter-row cage weeder.

Another example of non-powered cultivators is the star hoe or rotary harrow. It is ground-driven but has an aggressive weeding action. Disks can be adjusted to move soil away from, or in to, the crop row; the latter, ridging up the crop to bury small inter-row weeds. Increasing the forward speed of the rotary hoes does not improve the level of weed control. The machines work best on light, stone-free soils. The working depth is around 5 cm (2 in) at 8–12 km/h (5–7.5 mph).

One machine, the split hoe, incorporates two designs: a cutting movement with hoe blades and a brushing movement that uncovers uprooted weeds. It can be used on bigger weeds and in moister conditions than the standard hoe design.

Powered rotary cultivators these machines use cutting, uprooting and burial to kill the weeds. Typical rotovators are powered and normally comprise L-shaped blades fitted on to a horizontal axis that is usually covered (for safety and to prevent soil throw), although some use stout vertical tines that are rotated to stir the soil. They have an aggressive weeding action, pulverizing the soil, incorporating weeds and mixing the soil to a depth of around 12 cm (5 in). This intensive cultivation removes larger weeds and they are also used for seedbed preparation.

Brush hoes are primarily intended for inter-row weeding of vegetable crops, although they have been tested in cereals. As the name suggests, the

Fig 25 Rotary star hoes on rear mounted tool-bar.

weeding action comes from strong nylon brushes that rotate and brush the weeds on to the soil surface. They have the advantage that they can be operated under moister soil conditions than a steerage hoe (*see* above). A second person, in addition to the tractor driver, or some form of self-steering mechanism, is needed to ensure careful guidance of the brushes between the crop rows, or front mounting on the tractor may allow the driver to guide the brushes.

Two main types of brush hoe have been developed: those with disc brushes operating in the vertical plane on a horizontal axis, and those with circular brushes operating in the horizontal plane on a vertical axis. In the UK, only brushes that rotate on a horizontal axis are now available. Brush position on the drive shaft and brush width can be adjusted to different row widths, and crops can be protected by a tunnel that effectively guides the crop between the brushes.

In tests with the brush hoe on a horizontal axis, it was found that working depth was the most important factor in ensuring good weed control. Tractor speed, brush velocity and soil conditions interact to determine the working depth, and being able to adjust the depth precisely is a necessity. Tractor speed is limited by the operator's ability to steer the brushes close to the crop row at faster speeds (not a popular job in dry or dusty

Fig 26 Brush-weeder in action.

conditions!). At later weed growth stages it is necessary to increase the work intensity by, for example, increasing the brush speed relative to tractor speed, in order to maintain effectiveness.

Other powered cultivators. More recently a range of other powered cultivator-type machines have also been developed, including machines aimed at removing specific weeds (such as couch), machines using combinations of rotating tines and some using entirely novel rotating cutting blades (linked to computer-guidance systems). Many of these machines have yet to be extensively field tested and many are likely to be quite costly compared to more standard weeding equipment, at least in the short term. Some, however, promise, at least in some crops and situations, to improve efficiency of weeding operations.

INTRA-ROW WEEDERS

Removal of weeds in the crop row is the most difficult and delicate area in which to control weeds by mechanical methods due to the higher potential for harming the crop. Some level of intra-row weeding can be obtained with well set-up inter-row machines, such as the brush-weeders or rotary harrows. These will move soil into the crop row, covering and smothering the small weeds between the crop plants. There are, however, a growing number of machines available that are capable of intra-row weeding.

Fig 27 A rotary powered cultivator designed to remove couch.

Fig 28 A novel tine-weeder with a powered circular motion.

Finger weeders operate with two rubber discs of finger like projections, which are angled down and into the crop, one mounted either side of the row. These are attached to metal, ground driven spikes. These fingers work the intra-row soil, interlocking and moving between the crop plants. The crop needs to be well established to avoid the uprooting effect, so it is essential that there is a difference in size between the crop and weeds. The distance between the discs can be altered to allow a gentle or more aggressive weeding action. The machine works best on loose soils and is often mounted on a tool-bar behind a set of hoes that are used to break the soil crust.

Torsion weeders have a simple design comprising two spring tines angled backwards and downwards, either side of the crop row. The coiled base allows the tips to flex with the soil contours around established plants, uprooting small weeds (those in the white-thread stage) in the row. The position of the tines can be altered, depending on the level of aggression required.

Fig 29 Finger weeder mounted on specialist weeding machine.

Fig 30 A home-made torsion weeder.

Thermal Control

The two main methods of thermal weed control currently being used commercially are flame and infra-red weeding. Other sources of heat, especially steam and hot water, have also been used to kill weeds and some of these are also briefly described below. Thermal weeding can be used for total vegetation control or for selective removal of unwanted plants.

FLAME WEEDERS

Flame weeders use liquefied petroleum gas (LPG, propane) to fuel the burners. Weeds are killed by an intense wave of heat, from a direct flame, that ruptures the plant cells. Burners are generally mounted on tractor tool-bars or specialist rigs. Hand held flame weeders are available but these are generally used for weeding in amenity and industrial situations. Backpack flamers with an extended wand or an operator-pushed wheeled multi-burner are possibilities for use by small scale vegetable growers. It is a foliar contact treatment and any long-term effect depends on whether the injured plants recover and on the extent of subsequent weed emergence. For best effect, flaming requires a level soil surface and small weeds.

Fig 31 Burner with shrouded nozzles.

Plant size at treatment has a major influence on the level of control of individual species. Overall efficacy will depend on the species present. There can be problems when susceptible weeds are killed, only for the tolerant weeds to take advantage of the extra space. Grasses are known to be more resistant to flaming than broad-leaved weeds. Perennial grass weeds, such as couch, are likely to regrow rapidly after treatment. The more common annual weeds have been classed by their response to flaming as:

- sensitive species with unprotected growing points, e.g. fat hen and chickweed;
- moderately sensitive, requiring a higher energy dose, e.g. the knot-grasses (*Polygonum* spp.);
- flame tolerant that can only be killed completely at early growth stages, e.g. shepherd's purse or *Chamomilla suaveolens*;
- very tolerant species with a creeping habit or protected growth point, e.g. annual meadow grass.

The wide range of production methods used in horticultural crops offers more opportunities for using flame weeding than the arable crops. Vegetables are also relatively high value crops, where the cost of investing in flame weeding is justified. Selectivity may be achieved by timing the application to kill weed seedlings before the crop emerges (pre-emergence flaming). A sheet of glass laid along a short length of crop row can give advance notice of imminent crop emergence. Once the crop has emerged, angling or shielding the burners may allow selective inter-row weeding, or the burn may be adjusted to a level that the crop will tolerate (post-emergence flaming). Flame weeding is not suitable for crops with shallow or sensitive root systems. The flaming of weed seedlings prior to crop emergence should be delayed for as long as possible to ensure that the maximum number of emerging weeds are exposed to treatment. Flaming does not appear to reduce subsequent weed emergence and may even increase the germination of some species. However, unlike mechanical methods of weed control, there is no soil disturbance to stimulate a further flush of seedling emergence. Flame weeders have the advantage that they can be used when the soil is too wet for mechanical weeders.

INFRA-RED BURNERS

These are fuelled by propane/butane but the burners heat ceramic or metal surfaces that radiate the heat towards the target plants and kill the weeds by infra-red radiation. They are normally available as specialist tractor mounted rigs. Infra-red burners potentially provide a more reliable and constant heat source but have the disadvantage of needing time to heat up. In general, flame weeders are considered to provide higher temperatures, although temperature is not the only consideration as burner height and plant stage are important factors as well. The infra-red panels are sensitive to mechanical damage and the kit is more expensive than flame weeders. Infra-red burners can be used in similar ways to flame weeders (*see* above), although it is generally more difficult to direct the heat that is reflected off a large surface.

A hand held infra-red weeder is available that can be used for killing the rosettes of perennial broad-leaved weeds in grass. A ceramic disc, heated by gas from a small butane cylinder, generates infrared radiation when incandescent. The 'hot spear', as it is called, also has a projecting metal spike that heats up and this is pressed into the centre of the plant to be destroyed and held there for a few seconds. It has yet to be determined how effective the tool is against deep rooted weeds.

STEAMING AND HOT WATER TREATMENTS

Steam, generated by fixed boilers, has been used widely to sterilize the soil in glasshouses to control both weeds and plant diseases. However, in

Fig 32 A large tractor mounted infra-red burner.

organically certified systems, it is only allowed in protected structures as a one-off treatment to combat particular pest problems, and is not allowed for routine weed control. Notwithstanding this, mobile steaming equipment is now available to control pests, pathogens and weeds in polytunnels and in the field. Research has shown that steam forced down on to freshly formed beds for periods of 3–8min, raises the soil temperature to 70–100°C (158–212°F) killing most weed seeds to a depth of at least 10cm (4in). Low-temperature soil steaming for a short duration has been investigated as a more acceptable method of pest, disease and weed control and it has been shown that steaming soil samples at 50–80°C (122–176°F) for 3min kills fat hen and couch, as well as certain crop pathogens and nematodes. Ongoing trials of weeders that use an application hot foam to control weeds are also being undertaken. The foam is formed by a foaming agent in boiling water, which is directly applied on to the weed foliage through a nozzle. The foam has the effect of prolonging the high temperature of the water around the weed leaves for longer and, therefore, has an enhanced scorching or burning effect, over and above boiling water alone. It does not appear to affect below ground structures, as the water quickly loses its heat. All of these methods are currently restricted in organic systems and advice should be sought before using any of them.

FREEZING TREATMENTS

Low temperatures, as well as high ones, can destroy plant tissue. Liquid nitrogen and carbon dioxide snow (dry ice) have been evaluated for killing weeds but neither was as good as flaming. Freezing would only be advantageous where there was an obvious fire risk from flaming.

SOLARIZATION

This is a method of heating moist soil by covering with plastic sheeting to trap solar radiation. The covers can be left on for up to six weeks. Unlike steam sterilization, solarization does not sterilize the soil fully and create a biological vacuum, but there is some control of soil pathogens. For solarization to be effective it requires a climate with long periods of clear skies and sunshine to heat up the soil under the sheeting and maintain a temperature greater than 65°C (149°F) for long enough to kill the weed seeds present. Not all weed species are susceptible to high soil temperatures and even under ideal conditions the effective depth of control may be limited.

Research has been confined mainly to countries having suitable climates; there have been few investigations of solarization in Europe. In the cooler climate that prevails in the UK, studies of vegetable cropping under polyethylene sheeting indicate that weed development may be enhanced rather than impeded by the covers; but covers laid in midsummer could prove to have some limited weed control benefits. Disadvantages to using solarization for weed control include the loss of crop production for six to eight weeks in summer, as well as the purchase, laying and disposal costs of the plastic sheeting.

ELECTROCUTION

The concept of using electrical energy to control weeds was developed in the late-nineteenth century but more recently tractor mounted machinery was developed for controlling tall weeds that project above the crop. The safety aspects of using high-voltage electricity, which effectively flash heats the weeds, would necessitate operation of the machinery by a specialist contractor. Development work in the UK has been curtailed.

Mulching

Mulching provides a physical barrier over the soil surface through which light cannot penetrate. It prevents weed seed germination and suppresses seedling emergence. There are three main types: living mulches, particle mulches and sheeted mulches, which are described in more detail below. Mulches have long been used in soft fruit and in other perennial crops, and in situations where the treatment is intended to remain effective for many years. Mulches may also be used as an alternative to cultivation to clear vegetation before cropping. They are also more practical to use in well-spaced crops, particularly those that are transplanted, such as squashes (*Cucurbita* spp.). In freshly prepared seedbeds, short-term mulching can be used to manipulate or reduce weed seedling emergence. Black polyethylene is generally left down for the duration of a crop but studies have shown that sheeting laid on the seedbed for a short period, and then lifted before planting, can reduce subsequent weed emergence, giving the crop an advantage over the weeds.

Mulching is generally regarded as of high cost, especially where inputs like plastic or wood chips need to be purchased, and where large

Fig 33 Onions mulched with black polyethylene sheeting.

areas are to be covered. This only generally makes mulching economic for high value horticultural crops and for amenity use. Although mulches may be expensive, the reduced labour costs due to weeding can repay the investment over the long term. Set against this there can be practical problems with covering large areas for long periods.

LIVING MULCHES

These consist of a dense stand of low growing species established prior to, or after, the main crop. The undersowing of cereals with clover and grass can be seen as establishing a living mulch. Living mulches are sometimes referred to as cover crops, but they grow at least part of the time simultaneously with the crop, while cover crops are generally killed prior to crop establishment.

Often, the primary purpose of a living mulch is that of improving soil structure, aiding nutrition or avoiding pest attack. Weed suppression may be considered just an added benefit. In cereals, an understorey of clover has been shown to improve soil fertility, and reduce pest and disease problems, in addition to suppressing weeds. Maintaining vegetation cover is important for preventing soil erosion, nitrate leaching and weed emergence in crops that develop slowly. However, when the growth of a living mulch is not restricted, or when soil moisture is inadequate, even a relatively vigorous crop may suffer competition and loss of yield from the effects of the mulch itself.

Living mulches are well suited for use in perennial crops such as fruit, where self-reseeding is an advantage. However, even in established orchards, mulch growing along the planted row may depress crop growth and severely reduce marketable yield of fruit. Reduced growth of the crop may be due to competition for water or some other limited resource, or the mulch may be have an allelopathic effect. It is, therefore, important to make the correct choice of living mulch.

PARTICLE MULCHES

Where a layer of loose material, such as straw, bark, cut residues from a previous crop or composted municipal green waste, is spread over the soil surface, this can provide effective weed control but the depth of mulch needed to suppress weed emergence is likely to make transport costs prohibitive, unless the material is produced on farm or close by. A 3cm (1in) layer of compost is the minimum needed to prevent the emergence of annual weeds and control usually improves as the thickness of the organic mulch increases. Specialist baby leaf salad producers have used a thin layer of fine, evenly spread compost on beds used to produce the crop, in which there is no tolerance for the presence weed leaves.

Weed seeds in the mulch itself can be a problem if the composting process has not been fully effective or there is contamination by wind-blown seeds. In straw mulches, volunteer cereal seedlings are a particular problem due to shed cereal grains, and even whole ears, remaining in the straw after crop harvest. With particle mulches like straw, which consist of light materials, there is the possibility of them being blown around by the wind. Organic mulches with a high carbon to nitrogen ratio may deplete the soil of nitrogen as they decompose but this only occurs at the soil surface, unless the mulch is incorporated.

SHEETED MULCHES

These use a layer of material such as plastic, paper, woven plastic or even fabric to cover the soil surface. Black polyethylene mulches are widely used for weed control in organic vegetable production systems in the UK and elsewhere. Sheeted materials are relatively expensive and are often best laid by machine. Hand labour can be effective but can be fiddly and expensive. Heavy duty plastic is used for long-term crops such as perennial herbs. Woven polypropylene fabrics allow water to penetrate and are less likely to scorch crops when temperatures are high. Non-woven black fabric mulch may not be sufficiently opaque to prevent weed growth completely. After cropping, lifting and disposal may be a problem with plastic and other durable mulches.

Sheeting made from paper and other organic fibres have the advantage of breaking down naturally, and can be incorporated into the soil after use. Tearing and wind blowing can be a problem for paper mulches, but correct laying of the paper and rapid crop establishment are the keys to success. There can be additional environmental benefits if the paper mulch is made from recycled materials, such as cardboard. Crêping of some paper products has been tried in order to reduce the damage from stretching and contracting following wetting and drying.

A more recent development has been the increasingly widespread availability of biodegradable mulches. These are based on potato or corn starch and can be particularly effective in preventing weed growth during the establishment of crops such as courgettes (*Cucurbita pepo*) before they fully cover the ground. From an environmental point of view such biodegradable plastic is preferable but, by its nature, the mulch degrades and can cause unsightly wind-blown fragments around obstacles like hedgerows as it decays. Though the original outlay on material can be more expensive, there are no disposal costs associated with biodegradable plastic.

Although black polyethylene is the 'standard', a wide range of other materials have been developed that can selectively filter out different

Fig 34 Biodegradable mulch under salads.

wavelengths of radiation. One type can filter out the photosynthetically active radiation but let through infra-red light to warm the soil and these have been shown to be effective in controlling weeds. Various colours of woven and solid film plastics have been tested in the field. White and green coverings had little effect on the weeds, whilst brown, black, blue and white on black (double colour) films prevented weeds emerging. There are indications that mulching films, like white on black, with a higher rate of light reflectance are also beneficial to the crop. Light reflectance may also affect the behaviour of certain insects, and plastic mulches in a greater array of colours are likely to become available in the future.

Biological Control of Weeds

In its widest sense, biological control could be taken to include such basic practices as crop rotation, but the term biological control is usually restricted to the deliberate application or manipulation of natural control agents, normally a pest or disease, to control a specific weed. Biological control is a long-term strategy that requires a detailed knowledge of the ecology of the weeds and the natural enemies involved, and it usually requires a co-operative social effort to be successful. There are various ways of stimulating natural biological control of weeds, including inoculative control, inundative control and conservation control, and these are briefly discussed in more detail below. In addition to this, allelopathy can be regarded as a form of biological control, in which one plant is used to suppress the growth of another; as can broad-spectrum or total vegetation control, which, as the name implies, involves managing the whole habitat rather than a targeted weed species.

It is essential that any foreign biocontrol agents are thoroughly tested for host specificity to ensure that they do not pose a threat to other plant species and themselves become a problem! In a protected crop situation, an introduced biocontrol agent may remain contained because it will not survive outdoors in the UK, but it is much more difficult to control the likely spread and effects of agents released into an open field.

Classical Biological Control

Classical or inoculative control is the introduction of host-specific, exotic natural enemies to control alien weed species. For instance, many of the annual weed species in the UK have been introduced at some time in the past (alongside the crops in which they appear) and could be considered for classical biological control. However, since their arrival, often a long time in the past, most have become an established part of the flora and, as such, their wholesale destruction by exotic pests or diseases would not be welcomed. It has been suggested that some of the introduced, invasive perennial weeds, such as giant hogweed (*Heracleum mantegazzium*), Himalayan balsam (*Impatiens glandulifera*) and the Japanese knotweeds (*Reynoutria* spp.), would be ideal candidates for classical biological control. However, it is possible that these and other weeds may have some, as yet undiscovered, desirable feature.

The introduction of a classical biocontrol agent need not be deliberate. The groundsel rust (*Puccinia lagenophorae*) is of Australian origin, where it attacks a range of *Senecio* species. It was unknown in Europe before 1960 but since then it has been recorded in France and the UK on groundsel. The rust does not kill the weed but makes it less competitive. Higher yields have been recorded in lettuce experiments with rusted groundsel, compared with rust-free plants, indicating the types of effects that might be obtained. In practical terms, in the UK, the only candidate for classical biological control has been the perennial weed bracken (*Pteridium aquilinum*). Attempts to use the caterpillars of two species of South African moth as potential biocontrol agents have not, however, been successful.

Inundative Biological Control

Inundative or augmentative control involves the mass production and release of native natural enemies against native weeds and, as such, involves the culture and release of large numbers of a biological control agent into the region or field where the target weed needs to be controlled. It has the advantage that native organisms can be used but there is the same requirement for host specificity. Some agents, particularly plant pathogens, can be applied as sprays (often called mycoherbicides) in the same way as conventional herbicides. However, bioherbicides have the dual hurdles of the regulations that apply to biological control agents, as well as those that apply to a conventional pesticide, and none are currently available for use. Because of this, the acceptability of 'natural' herbicides to the organic standards authorities is also unclear.

Conservation Biological Control

Conservation control is an indirect method, whereby the natural level of the pests or diseases that attack the native insects that feed on the target weeds, are reduced and maintained at a low level in order to allow these herbivores to devour the weeds. Conservation biological control requires a

Fig 35 Dock beetle (Gastrophysa viridula) *larvae defoliating a dock plant.*

detailed ecological knowledge of the weeds and control agents involved. It has received little attention and remains largely a theoretical concept based on a reduction in the native parasites, predators and diseases that attack the native biological control agents of the target weeds. An example might be to reduce predation of the cinnabar moth caterpillar that feeds on common ragwort, but understanding the ramifications of this kind of system are probably beyond the ability of current ecological science and, given the current organic standards on biodiversity, probably undesirable in any case.

Broad Spectrum Biological Control

The oldest example of broad spectrum biological control, and possibly the most applicable to organic systems, is the use of grazing animals. Livestock have long been used to modify natural environments and can have a profound impact, especially in preventing regeneration of climax vegetation communities (for example, scrub and woodland in upland areas of the UK). It is known that different breeds of livestock vary in their grazing or browsing preferences and abilities, and this should be taken into account for improved weed control in pasture. In cereals, sheep grazing in spring is a traditional practice of many organic growers to aid weed control. Such concepts are discussed in more detail in Chapter 2.

Fig 36 Mixed grazing for broad-spectrum biological control of weeds.

Allelopathy

Allelopathy refers to the direct or indirect chemical effects of one plant on the germination, growth or development of neighbouring plants, and can be regarded as a form of biological control. The effect is exerted through the release of allelochemicals by the growing plant or its residues. Micro-organisms may also play a role in the production of these chemical inhibitors. Allelochemicals may be present in the mucilage around a germinating seed, in leachate from the aerial parts of plants, in exudate from plant roots, in volatile emissions from the growing plant and among decomposing plant residues. Both crops and weeds are capable of producing these compounds, which have, presumably, evolved, at least in part, to enable plants to reduce competition from nearby plants. From this idea it follows that allelopathy could be useful to organic growers in suppressing weeds. In the field, the evidence for allelopathy has largely come from studies of the use of organic mulches and cover crops to suppress weed emergence.

The effectiveness of living mulches, inter-crops or smother crops may in part depend on their allelopathic ability. The decomposition products of organic mulches and cover crop residues may continue to prove toxic to weeds in subsequent crops. Unfortunately, such phytotoxins are also known to reduce the germination and development of small seeded and drilled crops. Even the growth of transplanted crops may be checked.

From another angle, allelopathy could be used to manipulate the crop–weed balance by increasing the toxicity of the crop plants to the weeds. Where a crop has only a limited allelopathic effect, it may still be sufficient to reduce the emergence of difficult to control weeds in the crop row, leaving only the inter-row weeds to be controlled mechanically. Only in a limited range of crops have studies been made to evaluate the allelopathic ability of different cultivars or their tolerance to allelochemicals produced by weeds. Another approach has been to determine the crops that contain those chemicals, or their precursors, with the potential to suppress weeds. The glucosinolates, for example, precursors of several toxic metabolites, including isothiocynates, are found principally among members of the brassicas (Cruciferae). There have been suggestions that the allelochemicals themselves, or a synthetic derivative, could form the basis of 'natural' herbicides, although such an approach is probably contrary to organic principles.

Chapter 4

Economics of Weed Management

(*with Ulrich Schmutz*)

Costs will determine whether any particular farming enterprise is profitable or not and the costs associated with weeding are one of the key factors that can determine whether it is worth growing a crop. In this chapter, we discuss methods for estimating the costs attributable to weeding operations and introduce concepts for evaluating the costs as compared to the benefits. We also discuss weed management within a wider framework; that is, within the context of the farm system, which should also include an appreciation of the ecological and social consequences of any management operations or strategy.

COSTING WEEDING OPERATIONS

The cost associated with a weeding operation or weed management strategy is usually one of the principle considerations when planning and will often determine the type of action that will or can be taken. All farmers and growers are likely to be limited in at least some of the resources they need when it comes to weeding, but especially in time, labour and money. In this situation it is necessary to balance the likely benefits gained from weeding with the cost of carrying it out. Unfortunately, both costs and benefits are often subject to large uncertainties, not least because the likely efficacy of the operation itself (that is, how effective the control measure is likely to be at removing the weeds) will also be dependant on uncontrollable factors (like the weather or the tractor breaking down). Actual decisions are also made in a wider framework that will include a broad range of other ecological and social factors (like the market for the products, organic standards or even chance events like disease outbreak and changing government policies), all of which play a subtle part in deciding how and when to carry out weeding operations within a season and, therefore, what their costs are likely to be. However, it is still worth keeping records on the costs associated with weeding operations as it will, over time, show the profitability of any particular enterprise, but perhaps more importantly, indicate where limited resources might be better used.

In order to think about weeding costs it is necessary to identify the various cost components that go into a weeding operation, and to attribute costs to them. For costing purposes it is often convenient to give a cash value to all costs in order that different options may be directly compared, but it is not always easy to do this. For example, many ground preparation operations have both a weed management function and a seedbed preparation function and it will be difficult to share the costs between them in a meaningful way. In addition to this, organic farmers are often concerned to minimize the environmental impacts of their weeding operations and, in this case, it is often necessary to give a cash value to environmental effects that a weeding operation will have. This is doubly difficult as the cash value is often a matter of opinion, as well as of uncertain monetary value. For instance, what is the cost, in monetary terms, of disturbing ground-nesting birds? Sometimes it may be possible to give such an effect monetary value, say the value of an environmental management programme to farm income, but often this will not be the case. Indeed, many organic farmers are concerned about the environment for its own sake, an ethical judgement, and justifying this in cash terms will always be a difficult.

Elements of Weeding Costs

Despite the previous reservations, weeding operations will, on the whole, share the same basic costs and these are outlined below.

Labour

Most weeding operations will require some labour input, either skilled or semi-skilled. Manual labour (or hand weeding) is costly in terms of the amount of ground covered per unit time, as it is usually slow. If the weeds are hit at the right age, hand weeding is still a quite efficient and selective form of control, especially if the weeds are in the crop row and/or a crop needs rescuing. In some cases it is possible to speed up the process slightly by using machines like flat bed weeders that carry the workers over the crop and thereby force the pace, but in this case, the skilled labour of the tractor driver needs to be taken into account and the work will only progress at the pace of the slowest or be poorly done. Training or experience can also improve manual weeding effectiveness but this in itself might have a cost.

Mechanical weeders (*see* below) are more efficient in terms of work rate, although most require time for tractor drivers and may also require an additional person to guide steerage hoes when controlling weeds close to

Fig 37 Labour costs are a significant part of weeding costs; here a gang weeding onions.

the crop row. The time taken to set up machinery and prepare the operation can also be considerable, and many farmers and growers work on standard row or bed widths, minimizing time necessary for set-up. Routine maintenance can also be time consuming, as can carrying out any modification work necessary.

Some time should always also be set aside for management and design of rotations, management of personnel and other time such as for ordering spare parts, which are often not counted in, but can amount to a considerable cost, especially for smaller businesses or those with a limited number of personnel.

Machinery

Machinery can be expensive, both to buy and to maintain, but these costs are carried over the working life of the machine. Most farmers and growers will have at least one piece of tractor mounted weeding kit available, which they can use at short notice, and which they are comfortable with on their soil and in their farm situation. In many cases, farmers and growers modify kit that they have bought off the shelf and for this purpose, even cheap second-hand machinery can be more than adequate. In some cases, machinery that is only used occasionally is better hired in when needed, either from neighbours or from contractors or machinery rings. Apart from large pieces of kit, weeding may also require smaller hand-held equipment like hoes, billhooks or specialized pullers, which also have a finite (although potentially long) working life.

Additional Resources

Other resources might be needed for effective weed management and accounting for some of these can be important, especially as concerns wider ecological impacts. For instance, many machines use diesel and mineral oil lubricants. Flame weeding can use large amounts of liquid gas. Spare parts for maintenance to replace ordinary wear and tear can also be a factor in some machinery (for example, brushes on brush-weeders have a limited life compared to the machine frame). Plastic mulch is often used as a weed control method in vegetable production systems and, apart from the cost of the original material, there can also be costs associated with safe disposal, as well as the ecological cost of using plastic. In some cases, these costs can be overcome by using biodegradable plastic (starch) mulches, although the outlay costs may be greater.

Recording and Calculating Weeding Costs

In accounting for costs it is necessary to record them. This is usually best done at the time that the operation is carried out and a diary, log or register of all costs should be kept. This can be made more valuable by recording the impact of operations and making notes of what happened, and even ideas for future modifications. It is amazing how many good ideas are not acted upon because they are not recorded.

Hand Weeding

Hand weeding costs are usually simple to record. They are based on the hours worked and the rate per hour. Work can be for an hourly wage but, for weeding, is often allocated on a piece rate basis, that is, a fixed wage per area covered, irrespective of time taken, although this system is open to abuse on both sides. In this latter case, costs can be directly allocated per unit area covered but in the former, the area covered may need to be calculated. Because hand weeding is, in most cases, more expensive, it should only be used to clear the field from weeds not sufficiently controlled by mechanical weeding.

Mechanical Weeding

Measuring mechanical weeding costs is more complex because not only the work hours and rates for driving or steering have to be recorded, but also the costs of the tractor and the weeding machine itself, and its maintenance, taken account of. Here we describe a quick hands-on method for doing this on farm for costing and comparative purposes.

TRACTOR COSTS

Most farms have one or more tractors, which may, in some cases, be quite old. In order to calculate the costs of using them (Table 1) it is necessary to know the purchase price (second-hand or new), the depreciation rate, maintenance costs and the costs of fuel and oil used. Maintenance includes all repairs, insurance and taxes. The depreciation per year can be simply calculated over the projected lifetime of the tractor by dividing the difference between the buying price and the potential trade-in price by the projected use in years. Obviously any calculated depreciation figure can vary enormously between businesses depending on numerous factors (and is often set on advice from tax accountants), but it is usually prudent to assume a shorter, rather than unreasonably long, working lifetime for a tractor and to be realistic about any potential resale value.

Normally all costs are calculated as a cost per unit area (hectares or acres, *see* below) so that, for example, weeding operations can be compared across seasons, fields, farms or businesses; but tractors are used in many different operations and it makes more sense to calculate an hourly rate. The easiest way to do this is to divide the sum of depreciation (the lost value per year), maintenance (cost per year) and fuel/oil (cost per year) by the work rate in hours per year for the tractor. In the given examples (see spreadsheet in Table 1) a cost of £9.80/h for a new tractor and £7.20/h for an old one has been calculated based on 250h use per year. Weeding costs can then be calculated on the basis of the number of hours they take on an area basis (*see* below).

MACHINERY COSTS

A wide range of different pieces of weeding kit exists and all will have different costs associated with their use. Important cost factors for machinery include the width or number of rows covered, the working speed and any time lost at the headlands, all of which will directly affect the work rate, normally calculated as the work rate in hectares or acres per hour (*see* Table 2 for examples with a brush-weeder and spring tine). If it takes a long time to adjust machinery, then the time to do this should also be costed in as per cent time lost. In fact most farmers and growers will reduce this cost considerably by working with standard row spacings and/or bed widths. Alternatively, all these factors can be directly timed if accurate work records are kept for field operations. Then, as for tractors, depreciation and maintenance should be included in the costs, in this case

Table 1 Costing use of new and old tractors in weeding operations (as £/hour)

Tractor (new)				
Purchase price	£26,000 (new)			£/h
Depreciation	10 Years	35 per cent trade-in	250 h/year	6.8
Maintenance	300 £/year		250 h/year	1.2
Fuel and oil	900 ltr	0.51 £/ltr	250 h/year	1.8
TOTAL				9.8
Tractor (second-hand)				
Purchase price	£9000 (used)			£/h
Depreciation	15 Years	0 per cent trade-in	250 h/year	2.4
Maintenance	600 £/year		250 h/year	2.4
Fuel and oil	1,200 ltr 0.51 £/ltr		250 h/year	2.4
TOTAL				7.2

by dividing the area covered by the machine per year to obtain a cost per unit area covered (*see* Table 2).

OTHER WEEDING COSTS

If additional resources, over and above labour and machinery, are used to control weeds, then these costs should be directly recorded (in the cost spreadsheet). The cost of fuel in gas burners, for example, will be a significant cost in flame weeding and should be included. In the case of other weed management strategies, such as use of plastic mulches, the spreadsheet should be adapted to accommodate the costs associated with the method. Costs may include the costs of safely disposing of unwanted or used material, especially when using materials like plastic sheeting.

TOTAL WEEDING COSTS

In order to calculate the total cost of any weeding operation, the costs of using the tractor, the machine (or method) and the labour input, all need to be added together. This has been done in Table 2 for a brush-weeder and spring tine (assuming the same new tractor is used from Table 1). In the examples given, a total rate of £66/ha (£27/a) has been calculated for the brush-weeder and £18/ha (£7/a) for weeding with an old spring tine. The advantage of analysing costs in this way is immediately apparent, as the

Table 2 Example of calculating costs for using weed machinery (as £/unit area)

Brush-weeder					
Width, speed, time	1.8 m	3 km/h	10 per cent time lost		
Work rate (ha/h)	0.5 ha/h	1.2 (ac/h)			
Purchase price	£4,000 (new)			£/ha	(£/a)
Depreciation	10 years	20 per cent trade-in	40 ha/year	8.0	3.2
Maintenance	160 £/year		40 ha/year	4.0	1.6
Labour	1 skilled	10 £/h		20.6	8.3
	1 semi-skilled	6.5 £/h		13.4	5.4
Tractor		9.8 £/h		20.2	8.2
TOTAL				66.1	26.8
Old Spring Tine					
Width, speed (ha/h)	1.8 m	10 km/h	5 per cent time lost		
Work rate (ha/h)	1.7 ha/h	4.2 (ac/h)			
Purchase price	£200 (old)			£/ha	(£/a)
Depreciation	20 years	0 per cent trade-in	5 ha/year	2.0	0.8
Maintenance	24 £/year		5 ha/year	4.8	1.9
Labour	1 skilled	10 £/h		5.8	2.4
Tractor		9.8 £/h		5.7	2.3
TOTAL				18.4	7.4

comparative costs of the two machines can be gauged and the areas of highest cost in each method can be easily seen. It is also possible to evaluate the effect of different ways of doing things; for example, less time lost at headlands or increasing work rate. By knowing the individual costs, it can also be judged if contractors could do a more cost effective job, or if a co-operative ownership of machines could be beneficial, or indeed if buying a new one is likely to be worth it in the long run.

Fig 38 New weeding kit can be expensive to buy and knowing costs will help with decision making.

COSTS OF WEEDING IN ORGANIC CROPS

As already discussed, weeding costs have to be judged against higher potential marketable yields, and hence income gained. Independent weeding costs are often difficult to obtain, as weeding is usually only one small part of a farm's costs, and often not singled out amongst other operations. Although some research work has attempted to address this issue in organic agriculture, both through research plots and on-farm case studies, it is dogged by the often unrepresentative nature of research sites as compared to commercial farms and the difficulty of standardizing information gathered across farms in case studies. Notwithstanding this, good sources of costings can be found in various farm management handbooks (*see* Bibliography) that are based on data from commercial farms, as well as research projects. Here we only introduce on some of the highlights of this information.

Thresholds

Although most farmers and advisors would like to define specific thresholds for taking action against weeds, where the income gained would repay the cost of taking action, in practice, this has proved impossible to deliver for various reasons including:

- **Variable weed flora.** The weed flora is made up of a range of species, and different species will be problematic depending on the seasonal conditions in which the weeds are competing with the crop. That is individually, and in concert, weeds will have variable effects on crop yield from season to season, and it is therefore difficult to define the outcome of competition between crop and weeds, the damage caused and ultimately ascribe a monetary value to it.
- **Variable efficacy.** It is also difficult to define the effectiveness of taking action against weeds. This is especially true of organic systems, where the effectiveness of mechanical weeding can depend on timeliness, weeding skill, weed type, crop type and weather conditions. Under such circumstances, it is once again difficult to define a given outcome (that is, percentage weeds controlled) for a given action (for example, one pass with a steerage hoe), leading to uncertainties in costing.
- **Using rotational approach.** Organic farmers and growers rely on weed management through cultural methods. Many of these methods are best viewed as having an effect over the whole rotation cycle. So, for instance, it may be advantageous to suppress weeds in a cereal crop, where it is relatively easy to do so, in order to benefit subsequent vegetable crops, where it might be less easy or cost more to control weeds. Costs cannot therefore be easily ascribed to any one particular control action.

Notwithstanding the many attempts to model thresholds, especially for herbicide application in conventional systems, in practical terms, thresholds will come down to individual farmer or advisor experience and practice. For instance, many farmers tolerate a certain percentage of weeds in a crop based on their experience of the effect of the past levels of the weed on yield. Tolerance will also be a reflection of the value of the crop and the likely effectiveness of any weeding operation. Another major consideration for farmers and growers is the consequence of letting a weed population get out of hand, thereby building up problems for future crops, even if the effect on the current crop is likely to be small. In this case, the cost of taking action can be offset against better weed management in future crops and this, once again, highlights the difficulties of defining thresholds for weeding operations.

Organic Vegetables

The best information on costing exists for organic vegetable production. Results have shown that in organic vegetable crops, weeding costs can be on average around £800/ha (£324/a) for casual labour and £120/ha (£49/a) for mechanical weeding. However, it is also true to say, that a large variation in costs has been encountered in practice on farms. In a recent series of studies, costs ranged from near £0/ha (£0/a) for potatoes in one case, to £1,600/ha (£650/a) for hand weeding of carrots or onions in two others. In this same series of studies, whilst the overall costs of organic vegetable production were fairly stable across years and crops – measured in this case as a coefficient of variation (or cv) of 2 per cent – the total weeding costs (hand and mechanical weeding) were much more variable (with a cv of 22 per cent), that is ten times more variable. There was also a positive correlation between high weeding costs and crop net margins, which, in other words, means that high weeding costs often represent money well spent because higher marketable yields of high value vegetables 'pay' for the costs incurred. This relation, however, need not hold in the future, especially if organic premiums are reduced, and will also certainly not be true for less valuable, commodity type crops (*see* below).

Fig 39 Costing operations is important for improving weed management, even if kit is old or home-made.

Table 3 Hand-weeding costs (£/ha) in selected vegetable crops as a percentage of total output

Crop	Calabrese	Leeks	Carrots	Potatoes
Cost (£/ha)	136	468	2852	28
Per cent of total output	3	4	27	0.3

Hand weeding often can be a major cost in organic vegetable production. Some costs of hand weeding are presented on Table 3 from a study of organic farms over a four year period. In this series of observations, carrots were by far the most expensive crop for hand weeding, as a consequence of being directly drilled and a poor competitor against weeds. But even here, a well timed pre-emergence flame weeding combined with a stale seedbed can significantly reduce any hand weeding needed. Leeks generally required some hand labour, although the steerage hoe can provide good weed control under favourable conditions. In many instances, brassicas, such as calabrese, required very little hand weeding and many potato crops required no hand weeding at all. There was considerable variation between wet and dry years in crops such as leeks, where average hand-weeding costs (for the four farms) were £307/ha (£125/a) in the dry year of 2003, and £703/ha (£285/a) in the wet year of 2004. Soil type also had a large effect with costs at least 100 per cent greater on a peat soil type as compared to a sandy soil, where there is better weed kill from mechanical weeding equipment and lower inherent weed pressure.

The skill in producing profitable organic vegetables will lie in striking the balance between (saving or reducing) costs without affecting marketable yields adversely. In general, replacing hand labour with mechanical weeding has been found to be a good cost reducing strategy. In one study, made during a period of converting to organic vegetable production, hand-weeding costs were reduced over four year period by 21 per cent, while mechanical weeding costs increased by 39 per cent. However, because mechanical weeding is much cheaper, it was possible to decrease overall weeding costs by 14 per cent, without negative effects on marketable yields.

Organic Cereals

The difference in prices obtained between high value crops, like organic vegetables, and commodity crops, like cereals, will necessitate a different approach to weed control. Commodity crops are characterized by relatively low prices, are generally grown by many farms and include crops like organic (feed) cereals. Justification of weeding costs will depend on the price of the commodity, which may be volatile and difficult to predict.

In general, the additional value of the higher yield in the fully weeded commodity crop is the maximum weeding cost justifiable and this leads to an approach where the maximum weeding costs are calculated by estimating the yield level of a non-weeded and a fully weeded crop and costing the difference. In most commodities, in most seasons, this will be quite a low figure. A farmer workshop on organic weed control (in the UK) estimated that about £40/ha (£16/a) was the maximum amount that could be justifiably spent weeding organic cereals. This figure, depending on individual machinery costs, would only usually justify one or two mechanical weed control treatments. With lower yields and more weed competitive cereals, like oats, no weed treatment at all may be economically justified. However, as noted previously, rotational considerations aimed at preventing the build-up of undesirable weeds can trigger a decision to use a 'window of opportunity' in cereal crops to, for example, control perennial weeds, where operations might be easier and cheaper. In this case, the costs should therefore be spread over the rotation as a whole, rather than simply costed against the cereal crop.

Using the new tractor at 9.8 £/h (as described in Table 1) with a £9/h pay rate, a 12m wide floating bed coil-sprung harrow would cost 11 £/ha, if worked on 200 ha/year. A small harrow with 6m, used on only 50 ha/year would cost 22 £/ha. This can demonstrate that having a contractor, with a competitive price, on a smaller holding, would be best solution for one general weed strike, leaving the small harrow for one treatment in problem areas and incurring added costs of 33 £/ha – below the £40/ha justifiable spending level on weed control. Obviously, this is just an example, but it demonstrates how to approach costings and how it can be used as an aid to decision making.

Organic Grassland

Weeding costs in organic grass are the most difficult to assess. This is because weed levels are affected by the type and intensity of the grazing or cutting for conserved feedstuffs. All these management options (rotational grazing, mixed livestock grazing, silage cuts) have profound effects on the weed flora, although their main purpose is not weed control. It is, therefore, not justifiable to entirely allocate those costs to weed control. It is, however, possible to more accurately cost operations that are exclusively done for weed control purposes.

The main mechanical weed control operation in grass is topping the field or parts of it. This can be done with an old tractor and topper, at costs as low as £10/ha (£4/a). In some circumstances, hand held machines, such as strimmers or lawn mowers, can be used. For instance, on patches of weeds or around and under electric fences, and areas that need to be relatively weed- and grass-free to avoid short-circuits. In all these cases, costs can be calculated on the same basis as already described for mechanical weeding (*see* above).

Problem weeds like dock or ragwort often have to be tackled by hand, and here the main cost is labour. Specialist weeding tools, which can improve work rates and effectiveness, may also be more expensive and can cost about £100, although they do not suit all working styles. Specialist dock-control gangs exist, some through machinery rings, and many offer their services via the internet. There are online work rate predictors available to work out the hours needed, depending on the weed density, the number of plant extracted per minute and the hourly pay rate. As an example, for ragwort, with forty weeds per 100 m^2 (1000ft^2 or 0.01ha) and five plants extracted per minute, costs of more than £120/ha (£48/a) can be expected at a £6/h pay rate. Doing nothing and leaving the ragwort problem until the weed density has doubled, will obviously also almost double the weed control costs. Similar calculations can be made for docks. Although costs can sometimes seem high, it is also arguably possible to spread the costs over the rotation, especially if an efficient manual weeding operation reduces weed density over many seasons.

ESTIMATING BENEFITS OF WEEDS

Eradicating all weeds is not a practical goal in organic farming. It would not only dramatically increase weeding costs (as discussed above) and eliminate profitability; it would also deprive the organic farmer of any of the many beneficial effects of weeds. A low and manageable population of weeds can have many desirable effects in an organic farming system through increased biodiversity; for example, as habitat for natural predators or as a green manure to reduce nutrient leaching after a main crop is harvested. These effects are difficult to measure and difficult to put a monetary value to. However, this does not mean they are non-existent or not important. Table 4 lists the potential benefits that weeds might have in any situation and provides pointers as to how an estimation of their potential monetary value can be made.

At the present, the offsetting of the 'costs' of weeds against any benefits they might bring into the farm is necessarily more of an art than a science. It will not always be possible to make highly accurate estimations of the monetary value, but often, in costing terms, it is better to make at least

Table 4 Potential benefits of weeds and methods of estimating monetary value

Weed benefit	Monetary benefit
Increased agro-biodiversity with more species abundance.	An organic system benefit, rewarded by higher support schemes for organic agriculture.
Pest-control effects.	Higher yields. Savings in other pest control costs.
Green-manure effects.	Savings in green-manure seeds, improving soil fertility.
Decreased nitrogen leaching.	An organic-system benefit, no direct cash reward yet (but likely in future), potentially lower costs for any nitrogen inputs.
More biomass turnover and active soil flora and fauna.	An organic system benefit, no direct cash reward, improved long-term soil fertility.
Trapping and recovery of nutrients from deeper in the soil profile (deep rooted weeds).	Savings in nutrient inputs, possibly less manure and application costs or less time in fertility building.
Retaining of soil moisture.	Higher yields. Savings in irrigation costs.
Decreasing risk of surface soil erosion.	Soil erosion can damage crop yields and has long-term negative effects on soil fertility.
Protection from wind erosion and sand blasting of vulnerable early crops (e.g. carrots) on sandy soils.	Sand blasting can wipe out whole crop. Reseeding or total loss are the consequences.

some estimate rather than ignoring the benefits altogether. In this way, at least some of the benefits of weeds and their presence can be costed into farm operations, and judgements made about the benefits of having at least a managed weed presence.

ENVIRONMENTAL COSTS OF WEED MANAGEMENT

Weed management practices are an area of farm activity that can have a potentially high influence on the overall environmental impact of

Fig 40 Butterfly on knapweed, a potential benefit of having at least some weeds.

organic farming (*see* Benefits of Weeds above). Weed control operations are generally energy intensive, that is they substitute labour with non-renewable energy use, primarily fossil fuel; organic systems are no different to conventional systems in this regard. Tractors and weeding machines also have an embedded energy cost of manufacture, although, offset against this, many can potentially have a long working life.

Energy Use

Standardized data on direct and indirect energy use can be used to assess the difference between weeding in conventional and organic production systems. One study compared the total energy use (in MJ/ha) in organic winter wheat with that in conventional winter wheat and in this case showed that the energy consumption for weed control in organics was only 38 per cent of the conventional. This study assumed one pass with a harrow comb weeder in the organic wheat and one pass with a sprayer applying 2ltr of herbicide in the conventional wheat. Even if two passes with the harrow comb weeder were done in organics, which is often not economic, the total energy use would be below conventional. A closer look at the data indicates that this difference is almost entirely due to the energy value of the herbicide used (480MJ/ha for the 2ltr/ha herbicide), where the harrow comb weeding used 250MJ/ha and the herbicide spraying 170MJ/ha.

However, the use of flame weeding in carrot production does not produce such a favourable comparison for organic production, as flame weeding with gas consumes up to 6,500 MJ/ha. The energy needed for the tractor and flame weeding pass, some 200 MJ/ha, also needs to be added to this figure. In some cases, several passes with an inter-row weeder may also be needed, using 250MJ/ha for each. The total is almost twice as much as the comparable figure for herbicide treatment in conventional carrots, assuming 170MJ/ha for herbicide spraying and 3,400MJ/ha for the 12.5ltr/ha of carrot herbicides that would be applied.

The contrast between these two examples demonstrates that energy use in organic farming is not, in all cases, better than that in conventional farming, and will to some extent depend on the situation and circumstances. It should also be borne in mind that the overall efficiency of an operation needs also to be calculated by taking into account the output or yield of the system, which can be 10 to 40 per cent lower (depending on crop) in established organic systems. It is, therefore, necessary to continue to develop weeding systems that reduce energy use on organic farms. For instance, replacing flame weeding with other methods, developing more efficient flame weeders and/or combined systems with hoes and flame weeders, could all have potential benefits in reducing energy use in organic agriculture. There is also room to develop light and precise mechanical weeding machines to further reduce the energy necessary in weeding operations. Future technical developments could include direct solar-powered and vision-guided robots.

However, all 'high technology' solutions to weed control come with embedded energy costs – that is the energy cost of development and manufacture – and there is therefore great scope for developing novel rotations and management methods that manage weeds in an ecologically intelligent manner. The use of combinations of suppressive and less competitive crops, the use of green manures, and the selection of varieties with allelopathic potential are all areas where both farmers and researchers working together are more likely to produce viable and sustainable weed management protocols for organic farms. Even the humble horse may have a comeback if energy use and carbon footprinting continue to step up the global agenda. Luckily, horses were never abandoned fully by devoted farmers and growers, and modern machinery is available for them; however, it is currently too small a niche to make any larger impact.

Ecological or Carbon Footprint of Weeding

Environmental damage is becoming an important concern as human activities, especially agriculture, increasingly impact on natural ecological cycles and affect the way they function. This is doubly ironic as agriculture depends on ecological cycles to function and be productive. One measure of the impact of our activities on the environment is to calculate an 'ecological footprint', normally expressed as the land area necessary to support a person's resource consumption or sustain an activity (usually given in hectares or acres). It is often also measured as a carbon footprint (which is a good proxy measure of energy use) that represents the area necessary to offset the carbon (for example, by biomass growth) produced by any particular person or activity. When multiplied by the number of people on the planet, these footprints sum to give an estimate of how much area would be needed to support the population with a particular production and consumption level. Most measures of environmental footprints in developed economies, where certified organic agriculture is predominantly practised, indicate that we would need the land area of between three and four planets to support our current activities.

Of this footprint about 20–25 per cent is attributable to agriculture (including food production and distribution) and it should be pointed out that organic agriculture, as currently practised, will only go a short way to reducing this. Equally obviously, weeding and weed management will only represent a small part of the resources devoted to food production and distribution, but in order to reduce the footprint of the whole system, each part is important. In this case, farmers and growers should think about designing their systems to maximize the opportunities for weed management using rotational methods. That is, reducing the amount of active weed control with machinery in favour of a more ecological approach that uses plant competition and diversity to manage weeds. 'Biological control' using livestock can also make a valuable contribution to weed management on organic farms. Such a biological approach would aim to minimize the use of non-renewable resources, such as fossil fuels, plastics as mulch, or timber harvested from primary forests. Where possible, renewable resources should be used and maximum life obtained from those that are used. In this case, the mantra of the ecological movement is a good guide: reduce, re-use and recycle. Hopefully, by adhering to such principles, the long-term costs of weeding on organic farms can be reduced and the long-term sustainability of businesses assured. However, it should be acknowledged that there are no easy answers and that solutions will only be found by the combined efforts of farmers and researchers working together.

OUTLOOK

This chapter has discussed the importance of integrating weed management into farm costs and the importance of understanding the costs associated with various weed control operations, in order to be able to better manage them. However, in moving from specific costs to a wider economic view, environmental economics of weed management emerges as a promising area for future development, as agricultural systems increasingly need to tackle the problems associated with resource depletion (especially fossil fuels) and environmental degradation (pollution and global warming). The potential for a more ecological approach to weed management is an exciting possibility. It is likely to come about by combining the ecological knowledge of the detailed interactions between weeds and crops, between weeds and crop fauna (especially insects), and between weeds and ecological cycles (for example, nutrient cycling and soil biological activity) with the knowledge of farming as developed by farmers over generations, including rotational and crop combinations. The effect of combining this knowledge will be to create a learning environment in which farmers can decrease costs and increase environmental benefits both for agro-biodiversity and energy use efficiency, and, in the long term, guarantee the sustainability and viability of their businesses.

Chapter 5

The Weeds

We have defined weeds as plants growing in the wrong place at the wrong time and it is, therefore, possible for any plant to be considered a weed. Fortunately, only a more restricted percentage of plants are more commonly recognized as weeds, although in 1984 the Weed Science Society of America listed 1,934 species as current or potential weeds! In this chapter we describe the life habits of the weeds or weed types that organic farmers and growers are more likely to encounter. For the more common weeds or weed types we have indicated the sorts of management measures that have proved, or are likely to be, successful. Obviously it has not been possible to cover every potential weed species, or indeed every potential management method, in detail but rather the aim of the chapter is to illustrate how information about the life cycle or habits of weeds can be used to take a systematic approach to their management, while adhering to organic principles.

Weeds are variously grouped according to their lifespan, growth habit, time of emergence, means of dispersal, biological origin and plant family, the crops they occur in and so on. In this book we have separated weeds into annuals, biennials, stationary perennials and creeping perennials, more or less based on their ecological survival strategies (see Chapter 1) because this would seem to be the most useful way of thinking about their management in organic systems. It should, however, be recognized that there are species that could fit into more than one of these categories. Another common way of classifying weeds is into broad-leaved and grass weeds, and this is often more useful where herbicides are to be used. Although this is not an option in organic farming systems, we have still maintained this distinction, especially between annual broad-leaved weeds and annual grasses. Even under this dual classification, species such as field horsetail (*Equisetum arvense*) and bracken (*Pteridium aquilinum*) do not easily fit into any categories and it is clear that, whatever system of classification is used, there will be weeds that do not fit into the chosen classes. Volunteer weeds that are derived from crops are a case in point, and in this chapter we have treated these as a separate class.

ANNUAL WEEDS

Annual weeds are so called because they complete their life cycle in on year. They originate almost exclusively from seed and the majority aris from seeds that remain in the soil from a previous season. A few will aris from seeds bought in by other means; some might develop from seed brought in as contaminants in crop seed, some in soil amendments lik manure and slurry, others in organic materials such as straw used fo mulching, some may be blown in on the wind, others may be carried in b birds and animals or on farm machinery.

Annual weed species are sometimes classified according to the time c year they emerge. The seedlings of some species will emerge any time tha conditions are favourable but others tend to emerge only at certain time of year. Weeds that have a restricted period of emergence often becom associated with the crops that are sown or planted at the time they ar likely to emerge. Weed species that germinate and emerge during sprin and early summer, then die later in the same year after flowering an setting seed, are termed summer annuals. Seedlings of summer annual that emerge late in the year do not usually survive the winter. Weeds tha germinate in late summer or autumn, and survive the winter to flowe and set seed in the following year, are termed winter annuals. Howeve some winter annuals will germinate in the spring too and can then act a summer annuals. Opinions may differ on the category to which wee species should belong. The response of species may vary from country t country due to the biotypes present and to climatic differences. It shoul be borne in mind that rainfall pattern and moisture availability are ver important factors in the timing of weed emergence and may over-rid other factors.

Managing Annual Weeds

The key to managing annual weeds lies in understanding the soil seec bank and, to a lesser extent, taking precautions with farm hygiene. In mos soils there is a reservoir of seeds that reflects the cropping and weec management history of the land, not just in the previous year, but ove many years. Most of this seed is from weeds (and crops) that have bee allowed to mature and drop seed, although small amounts can arrive b other routes, as described above. The soil seedbank, therefore, comprise a wide range of crop and weed species, but usually the seeds of only a fev species are present in large numbers. Seed numbers in the upper 15cı (6in) of soil in commercial vegetable fields sampled in the 1960s, before th widespread introduction of herbicides, had a median value of 100 million seeds per hectare (247 million per acre)!

Once weed seeds are mixed in the soil they become part of the weed seedbank; their persistence depends on the species, the depth of burial and the frequency of cultivation. The seeds of grasses tend to be less persistent than the seeds of broad-leaved annual weeds. Persistence is less when the seeds remain in the surface layers of a frequently cultivated soil. Other factors responsible for seed losses in soil include germination, predation and decay. Seed predators include insects, like ground beetles and other arthropods, as well as birds. Seeds are also subject to attack by soil microorganisms, like fungi, and once they die, will quickly be subject to normal soil decomposition processes.

Seeds in the soil seedbank can be stimulated to germinate by bringing them to the surface and exposing them to light or other favourable conditions. The ability of seeds to germinate is often limited by seed dormancy. In effect, dormancy prevents seeds germinating until it is broken. Several types of dormancy exist, which are influenced both by environmental (external) and genetic (innate) factors. Dormancy can be terminated by many cues including presence of nitrogen, low temperatures or fluctuating temperatures. Once dormancy is broken, the seed can germinate, although if the conditions are inadequate, a seed may then enter a secondary dormancy. Dormancy is an ecological adaptation that has arisen to spread risk over time and to prevent all the seeds of a particular species germinating simultaneously and potentially all succumbing to the same lethal growing conditions (such as drought or even weeding!). Any weeding strategy should take it into account as it is a major factor in determining weed seed persistence in the seedbank.

Under normal agronomic conditions, most soils lose 3–6 per cent of the weed seedbank through seedling emergence each year, but seed losses from other factors (predation, decay) are much greater. Cultivation often encourages germination, as more seeds are exposed to the right conditions. Seed numbers in soil can be reduced by 30–60 per cent per year by frequent cultivations, if there is no further seeding. Nevertheless it will take several years for seed numbers to reach a very low level. Even then a small number of highly persistent seeds will continue to remain viable in soil for a very long time. If weed control is neglected at any time, this seed reserve has the potential to replenish the seedbank again within a year. A dense stand of weeds can produce over one million seeds per m^2 (10 million/ft^2) if left uncontrolled for a single season.

It is vital that only clean crop seed is sown to prevent introducing weeds into the crop after cultivation. Stale or false seedbeds and delayed drilling are effective ways of reducing annual weed numbers before the crop becomes established. The conditions must be favourable for weed seed germination and for controlling the early flush of weeds by shallow tillage or flame weeding. The seedlings need to be small for flame weeding to be effective but annual grass weeds are not susceptible, even at early growth stages.

Where weed control is limited to just one or two operations, timing is important. The aim is to allow as many weeds as possible to emerge but then remove them before they begin to have an effect on crop yield. In many broad-leaved crops, the optimum weeding time can be around four weeks after crop emergence. In a competitive crop, the developing crop canopy will help to suppress further weed development and this is reflected in reduced seed numbers. In an open or poorly competitive crop, further weeding will be needed, usually by surface cultivations with mechanical weeders. With tine weeders, tap rooted weeds need to be controlled early before the seedlings become well established. Fibrous rooted weeds can be left longer to allow sufficient foliage to develop to catch easily on tines or harrows. The time of weeding is less critical with a hoe

Fig 41 Annual weeds are susceptible to mechanical weed-control methods; here a ducksfoot hoe.

blade that can undercut established weeds. Cutting at, or below, the soil surface with soil burial are the most effective means of mechanical control.

To reduce future annual weed problems it is important to limit or prevent weed seed production. Every opportunity should be taken to remove weeds before they flower. Weed seed collected by the combine harvester during cereal harvest, should be retained and denatured, rather than dropped back into the field.

After crop harvest, the timing and depth of post-harvest cultivations will influence the distribution of freshly shed seeds within the soil profile and this can have a major effect on seed persistence and subsequent weed emergence, depending on the species involved. For example, shed seeds of oilseed rape should be left on the soil surface to germinate, while seeds of barren brome should be buried as soon as possible after shedding to prevent them becoming dormant.

Broad-Leaved Annual Weeds

Annual broad-leaved weeds can be very persistent in the weed seedbank but are generally very susceptible to cultivation and mechanical weed control methods. It is vital to prevent the return of seeds to the weed seedbank. There are a large number of annual broad-leaved weeds, many of which will be over familiar to farmers and growers. We have highlighted the most commonly encountered, and used them to describe the range of situations that are likely to be found in practice and to illustrate the main points to consider in managing them. Further information can be found in numerous internet based factsheets, which can be a good source of information on many annual weeds and their management. When searching for sources of information, it is better to use Latin names in preference to common names, which differ between countries and localities. Below we have used the most commonly accepted common names in the UK and supply the full Latin names in brackets.

Black Bindweed (*Fallopia convolvulus* (L.) A. Love)

Black bindweed is a summer annual, common on light sandy soils, loams and clay. The weed twines around and drags down both cereal and root crops. It flowers from July to August, sometimes into October. Before the flowers develop, it is often mistaken for field bindweed (*Convolvulus arvensis*). All stages of flowering can be found on a mature plant, from buds to ripe seeds. A plant normally has 140 to 200 seeds but a large plant can produce 11,900 seeds. Seed is shed from August onwards.

The seeds have an impermeable coat and are normally dormant when mature. Scarification promotes germination but light has no apparent effect. In the field, most seedlings emerge from March to May. Black bindweed seedlings can emerge from as deep as 19cm (7in) in soil but the majority come from the upper 6cm (2in). Emerged seedlings represent around 8 per cent of the seeds in the seedbank. Seed longevity in soil is usually given as four to five years, but seeds buried for over thirty years are reported to have germinated. In cultivated soil, the seeds have an annual decline rate of 40 per cent.

Black bindweed seeds were a frequent contaminant of cereal grain and this is thought to have been a major factor in its spread around the world. The seeds are considered a valuable stock feed and some farmers collect up the cereal screenings and feed them to livestock. If the seeds are not denatured, viable seeds could be returned to the field. When cattle are fed the seeds, digestion causes a gradual decline in viability but viable seeds can be found in manure. Seed does not germinate after ensilage followed by rumen digestion. Black bindweed seeds are killed by two weeks windrow composting at 50–65°C (83.8–149°F).

Black Nightshade (Solanum nigrum L.)

Black nightshade is a troublesome summer weed. It occurs on a wide range of soils but prefers those rich in nitrogen. Two introduced 'green' nightshades also occur locally in some areas; they have a similar growth habit to black nightshade but the berries remain green, even when ripe. Black nightshade berries are a particular problem in crops grown for processing, especially vining peas. The berries are considered toxic and have caused varying degrees of poisoning in humans, cattle, pigs, goats, ducks and chickens. The evidence is conflicting, however, and consumption of the ripe berries does not always result in ill-effects. The plant also contains a high level of nitrate that may prove toxic to livestock.

Black nightshade flowers from July to September. A plant can produce up to 400 berries, each containing about forty seeds. An average plant produces 9,000 seeds but a large plant may have 153,000. The seeds from unripe berries have given 20 per cent germination. In the field, seedlings emerge from May to September, with a peak in June. Most seedlings emerge from the surface 2.5cm (1in) of soil. The seedlings and mature plants are susceptible to frost and late germinating seedlings are unlikely to reach maturity.

In cultivated soil, the seeds will remain viable for at least five years. The annual decline is estimated at 37 per cent. Seeds buried in undisturbed soil have given 80 per cent germination after thirty-nine years. Viable seeds have been found in bird droppings. Small mammals also disperse the fruits and seeds. Black nightshade seeds can remain viable in silage and survive digestion by cattle.

Charlock (*Sinapis arvensis* L.)

Charlock was formerly the most troublesome annual weed of arable land. It is more of a problem in spring-sown than autumn-sown crops. It is frequent on heavy soil, but is also found on sand, chalk and light loam. Charlock generally flowers from May to July, but flowering can begin as early as April in plants that germinated the previous autumn. Successive flushes of seedlings mature and flower through the summer. Charlock has ten to eighteen seeds per seedpod and produces 2,000 to 4,000 seeds per plant.

Some seeds will germinate at once, others remain dormant for long periods and germination is sporadic. Scarification of the seed coat increases the level of germination. In the field, seedling emergence takes place from November to July, with a peak in March to April. Rainfall and soil temperature have a big influence on the timing of emergence. Seedlings that emerge in the autumn are often killed by frost. Field seedlings emerge from the top 7cm (3in) of soil, with most seedlings emerging from the upper 4–5cm (2in). Charlock seeds buried in uncultivated soil can remain dormant for at least sixty years and will germinate when brought to the surface by deep ploughing. In cultivated soil, seeds have a mean annual decline rate of 23 per cent and an estimated time to 95 per cent loss of eleven to fourteen years.

A dense autumn or spring cereal crop will reduce seed production by charlock, but some cultivars are more competitive than others. Charlock is less competitive in dry conditions. The seedpods usually remain intact until crop harvest. In cereals, some seeds are gathered with the crop, but many pods shatter and the seeds fall to the ground. Stubble cleaning is an effective way of dealing with freshly shed charlock seeds by cultivating the surface soil to no deeper than 5cm (2in), repeated at fourteen-day intervals. Early harrowing of soil in preparation for a root crop will induce charlock germination, allowing the mechanical destruction of seedlings during subsequent seedbed preparations. Where land is infested with charlock, the soil may be cultivated at regular intervals to stimulate and kill successive flushes of charlock seedlings.

The seedpods and seeds are poisonous to livestock but the rest of the plant is harmless. Charlock seeds that contaminate oilseed rape at harvest will increase the linolenic and erucic acid levels in extracted oil. The seeds can also occur in oil cake with serious consequences for horses and cattle. Charlock seeds have been a frequent contaminant of cereal grain. Charlock is a host of the turnip fly, turnip-gall weevil, turnip flea beetle, cabbage root fly, diamond back moth and leather jackets, and is also susceptible to club root disease with potential serious consequences for these brassica crops.

Fig 42 Charlock in organic cereal.

Cleavers (*Galium aparine* L.)

Cleavers occurs on most soils throughout the UK but prefers nutrient rich sites. It is a frequent weed of winter cereals, where it forms dense patches that grow over and drag the crop down, making harvesting difficult. Cleavers flowers from June to August and the average number of seeds per plant is 300 to 400. The fruits are covered with hooked bristles that catch on clothes and on animal fur.

Germination is inhibited by light, and seeds do not germinate unless covered with soil. The optimum germination temperature varies with seed age but in the field, germination usually ceases above 15°C (59°F). Seedlings emerge from August to May and are resistant to frost. In hedgerows, cleavers seedlings emerge in the autumn but, in arable fields, peak emergence is often from March to April. Maximum depth of seedling emergence ranges between 4 and 20cm (1½–8in) but most emergence is from depths of 2–5cm (1–2in).

Fig 43 Cleavers in wheat can complicate harvesting.

In cultivated soil, cleavers seed has been estimated to take about four years to decline by 99 per cent (or about 66 per cent per year), although emerged seedlings represent just 2 per cent of the seedbank. In dry storage, seed longevity is four to five years. Cleavers seeds have been frequent contaminants in crop seeds, especially home saved cereals, and farmyard manure may also contain seeds from contaminated straw used as bedding. Seed loses viability after thirty-four days in stored manure, and seeds in compost are killed when temperatures exceed 50°C (122°F) for two weeks. The seeds survive passage through cattle, horses, pigs, goats and birds. Seedlings have also been raised from bird droppings. The seeds will float in water but in flooded soil they lose viability after twenty days.

Around 60 per cent of freshly shed seeds will germinate if the surface soil is cultivated repeatedly. Buried seeds may also germinate following deeper cultivations, but will then fail to emerge, so that ploughing does not result in the persistence of cleavers seeds in soil. Minimum tillage, however, puts the seeds at the optimum depth for germination and emergence. Seed numbers in soil are reduced to less than 20 per cent by fallowing for a year. Cleavers is favoured by winter cropping. In cereals, harrowing with a tine weeder at an early crop stage can give a 79 per cent reduction in weed density. A second harrowing at a later crop stage improves the level of control. The growth of cleavers in headlands can be suppressed by sowing grass and wildflower mixtures.

Common Chickweed (*Stellaria media* L.)

Chickweed is one of the commonest weeds of cultivated land in the UK. It is widely distributed, but is more abundant on lighter soils and is absent from the most acidic soils. It thrives in areas of soil disturbance and declines when cultivation ceases for a long period. It grows best in cool, humid conditions but is sensitive to drought, and is one of the first weeds to wilt in dry conditions. The weed is an important constituent in the diet of many farmland birds.

Seeds will germinate at any time of year but particularly in spring and autumn. Some seeds can germinate immediately after shedding but buried seeds develop a light requirement. In the field, almost all seedlings emerge from the upper 30mm of soil. Common chickweed can complete its life cycle in five to six weeks. It is able to grow at relatively low temperatures and seedlings can survive all but the severest frosts.

Chickweed can be very variable in size, habit and general appearance. Summer and winter forms with different growth habits are thought to occur. It flowers and sets seed all through the year, even under a snow-cover. The flowers are normally self-pollinated but there are short periods

Fig 44 Chickweed seedlings can germinate throughout the year.

when insects can effect cross-pollination and, in winter, flowers are produced that do not open, making self-pollination inevitable. Plants hoed while in flower do not produce viable seed, unless immature capsules are present that continue to ripen. Each capsule contains around ten seeds and the average seed number per plant is 2,200 to 2,700. However, plants with 25,000 seeds have been recorded.

Buried seeds are known to retain viability for at least twenty-five years, and seeds in dry storage for thirty months at low temperatures have retained full viability. In cultivated soil, seed declines at an annual rate of 30–41 per cent with an estimated time of seven to eight years to decline by 95 per cent. Seed numbers in soil may be reduced by 85 per cent following a one year fallow. The mean annual decline under a grass sward is less at 26 per cent. The seed capsule splits when mature and the seeds fall on to the soil beneath the parent. Chickweed seed has been a contaminant in cereal, grass, clover and other crop seeds, and is dispersed further in mud on footwear and tyres. Seeds survive passage through the digestive systems of birds and may germinate in their droppings. Seeds are also found in cattle, deer, horse and pig droppings and in wormcast soil. Seed has been recovered from irrigation water.

In cool, wet conditions, common chickweed forms a dense mat of spreading stems that may root at the nodes making it difficult to hoe or pull up. Hoed plants will root again in moist soil. In root crops, control is by repeated surface tillage in hot, dry weather. In cereals, spring tine harrowing in July is said to give good control of the weed. On newly sown leys, grazing by sheep may to help to suppress chickweed, but mowing is ineffective and may help the weed by removing the taller vegetation. Chickweed seedlings with two to six leaves are relatively susceptible to flame weeding, and the seeds are killed by soil solarization. The seeds of chickweed are consumed by several species of ground beetle. The fungus *Peronospora media* may be an important agent in the natural control of chickweed. However, chickweed is a host of several damaging virus diseases of crop plants and these can be carried in the seeds and grow into infected plants.

Common Fumitory (*Fumaria officinalis* L.)

Fumitory is found on cultivated soils throughout the UK. It flowers from May to October and is self-fertile. Seed numbers per plant range from 300 to 1,600 with an average of 800. Light does not stimulate seed germination. In the field, most seedlings emerge from September to June, with the main flush from February to May and a smaller one in October to November. Seedlings that emerge in the autumn continue to grow through the winter, even at low temperatures. Field seedlings emerge from the top 9cm (3.5in) of soil, with the majority of seedlings emerging from between 5 and 6cm (around 2in) deep. Very few seedlings come from the surface 0–0.5cm of soil.

Seeds in undisturbed soil declined by 70 per cent after six years, and in cultivated soil this increased to 90 per cent. Seeds followed over a six year period of cropping with winter or spring wheat had a mean annual decline rate of 21 per cent. The estimated time to 95 per cent decline was ten to twenty-one years. Under a grass sward, common fumitory had a mean annual decline of 1 per cent and a half-life in excess of twenty years.

Fumitory seeds have been a contaminant in cereal and clover seed, and have been found in cattle droppings. In winter cereals, common fumitory is favoured by ploughing and discouraged by minimal cultivation. Fumitory is eaten by cattle and sheep, but horses avoid it.

Corn Spurrey (*Spergula arvensis* L.)

Corn spurrey is troublesome on light, sandy soils, deficient in lime. It occurs primarily on arable land, disturbed grassland and waste places, where it grows as a summer annual and is common in cereals, particularly spring barley. Corn spurrey flowers from June to August and its flowers

Fig 45 Fumitory seedling in brassicas.

are mainly self-pollinated, although insect pollination can occur. It produces seeds in summer and early autumn, until killed by frost. Seed starts to become viable one to two weeks after flowering and there are twenty-five seeds per capsule. The average seed number per plant is 1,300 but a large plant may produce 22,000 seeds. Plants can complete their life cycle within ten weeks and there may be two generations in a year. Two types of seed are produced that differ in the presence or absence of papillae on the surface of the seed coat but individual plants bear only one type of seed. The non-papillate seeds germinate more readily in moist conditions and at lower temperatures.

Seed dormancy is broken by rising temperatures in spring. Emergence occurs from February to October, with most seedlings appearing from March to July. Field seedlings emerge from the top 3cm (1in) of soil, with the majority coming from the upper 1cm (0.4in). Seeds can remain viable in soil for at least five years and over fifteen years has been recorded in dry storage. Corn spurrey seeds have been a contaminant of grass, clover and cereal seeds. Seed dispersal is possible in mud on the tyres of farm vehicles. The seeds have been found in cattle, horse, pig and bird droppings.

Fig 46 Small corn spurrey plant.

Corn spurrey is eaten avidly by many animals, particularly sheep, and has been included in grass seed mixtures. Badly infested crops may be grazed with sheep. Corn spurrey does not tolerate trampling. In grassland and set-aside, cutting time can influence the amount of seed returned to the soil. However, plants defoliated when in active growth can produce new shoots from the base. Control has been achieved by thorough liming or chalking.

Fat Hen (*Chenopodium album* L.)
Fat hen is a native summer annual found on cultivated land and in waste places throughout Britain, although it is less frequent in the north and west. It is common on sandy loams and clay, but less numerous on calcareous soils and gravel. It grows best on fertile soils and is especially plentiful (or obvious) in potatoes, sugar beet and other root crops. Fat hen is a very variable plant that shows wide morphological plasticity in response to soil fertility and plant density. It is known to hybridize with related species but the hybrids are difficult to identify due to the variability of the main species.

Fat hen flowers between July and September. It is wind pollinated and the flowers may be cross- or self-pollinated. Seed numbers per plant range from just ten to over 164,000. The seeds mature relatively late in the season. Fat hen produces several different types of seed on the same plant. Most seed is black and hard coated with either a rough or smooth surface. Up to 5 per cent of seeds are relatively large and brown with thinner, usually smooth, seed coats. The brown seeds germinate readily, while the black seeds do not and persist longer in soil. Damage to the seed coat of the black seeds will encourage more rapid germination if conditions are favourable. Plants that grow from brown seeds produce the same proportions of black and brown seed as those from black seed. Immature seeds are capable of germination and may do so more readily than ripe seeds, due to a thinner seed coat.

In the field, seedling emergence takes place from March to November, with the main flush from May to July. The number of seedlings emerging is positively related to the frequency of soil cultivation. Most seedlings emerge from the surface 3cm (1in) of soil with the odd seedling from lower down (6cm/2.5in). However, there is a marked decline in emergence from seeds on or near the soil surface. Plants that emerge earlier in the year tend to be larger and leafier than those that develop later. Fat hen is killed by frost, and seedlings that emerge in the autumn rarely survive the winter. Late spring frosts can affect seedlings that emerge early in the year.

Seeds buried in undisturbed soil for thirty-nine years have germinated and seed longevity in dry storage is eight to ten years. In cultivated soil,

Fig 47 Small fat-hen plant (among other weeds) in onion row.

the estimated time to 95 per cent loss of seed is six to twenty years, depending on the frequency of cultivation. There is no obvious seed dispersal mechanism. Fat hen seed was a common impurity in commercial clover, vegetable and cereal seeds, especially home saved seeds. Seeds have been found in cattle, horse, pig and bird droppings. Fat hen was the most numerous seed to survive in fresh organic dairy farm manure but numbers were reduced by 90 per cent after composting. The introduction of fat hen seeds with manure should be guarded against. Some seeds can survive ensilage or rumen digestion alone, but most seeds are killed by a combination of eight weeks ensilage followed by rumen digestion. Fat hen seeds have been recovered from wormcasts. Seeds have been found in irrigation water and are known to survive a long period of submergence in water.

Control in cereals is by surface cultivations with light harrows when the crop is 5–7.5cm (2–3in) tall. Large fat hen plants may need to be removed by hand to prevent seeding. In newly sown grass, fat hen seedlings do not survive cutting or trampling. Repeated cultivation is unlikely to deplete the seedbank of fat hen seed due to seed dormancy. Seedlings with two to six leaves are killed by flame weeding. Fat hen seed is susceptible to soil solarization and can be killed by direct heating of soil.

Field Penny-Cress (*Thlaspi arvense* L.)

Field penny-cress is an annual or over-wintering weed of arable land, roadsides and waste places. It is scattered throughout most of the UK.

It succeeds in both dry and moist habitats and thrives in nutrient rich loams. Seedlings that emerge in spring flower within thirty to fifty days and flowering is hastened by increasing temperatures. Penny-cress flowers from May to October and the first seeds are produced in early July, being subsequently shed over several weeks. Each seedpod contains around sixteen seeds and a plant may produce 1,600 to 15,000 seeds. Penny-cress plants smell strongly when crushed and can taint milk if eaten by cows.

Seeds can remain dormant for long periods but fresh seed is said to germinate in the light with adequate moisture and alternating temperatures. Seed dormancy can be relieved by chilling. Nitrate will stimulate germination but only in the presence of light, and scarification of the seed coat also promotes germination. In the field, seedlings emerge from February to October, with the main flush from February to June. Seedlings emerge from the top 5cm (2in) of soil, with most coming from the upper 3cm (1in). Seedlings that emerge in the autumn, over-winter as vegetative rosettes and flower the following spring.

Field penny-cress seeds buried in undisturbed soil have given 87 per cent germination after ten years and a small number of seeds have remained viable even after thirty years burial. In cultivated soil, survival is much less and few seeds survive longer than six years. Seeds will float for twenty-four hours and dispersal by water is possible. Viable seeds have been found in bird droppings and cow manure. In cattle, rumen digestion leads to a gradual loss of viability with time. Ensilage for eight weeks appears to kill field penny-cress seed. The seeds are also killed by windrow composting for two weeks at 50–65°C (83.8–149°F).

Gallant Soldiers (*Galinsoga parviflora* Cav.)

Gallant soldiers is an introduced weed of cultivated and waste ground. It was brought to Kew Gardens from Peru in 1796 and soon became common in the local area; there may have been other introductions of the weed. It increased rapidly in parts of the southern counties, where horticultural crops were grown. The closely related, but less common, shaggy soldiers (*G. quadriradiata*) is similar in appearance and occurs in the same situations.

Gallant soldiers flowers from May to October or until killed by frost. It can cross or self-fertilize and may have three to four generations in a year. The flower head has three to eight ray flowers and fifteen to fifty disc flowers. There are around twenty-six seeds in a flower head. The average number of seeds per plant is 2,000 but a large plant can have up to 15,000 seeds, and a figure of 400,000 has been suggested. The seeds are shed eleven to fourteen days after flowering and can be rayed (winged) or disced (pappus of hairs) as an aid to wind dispersal. In the UK, the main period of seedling emergence is from March to October. Light is needed for germination and soil burial induces dormancy in the seeds. Seeds germinate from the surface 2cm (1in) of soil. Seedlings are frost sensitive. The seeds of gallant soldiers remain viable in soil for two to five years. The ray seeds may persist at a low level beyond this time.

Seedlings can come into flower very quickly and should be hoed off when small. With little seed dormancy, regular cultivations can eliminate gallant soldiers in three to four years, if there is no further seeding. Putting land down to grass for a similar period has also proved effective. A thick layer of organic mulch will prevent seedling emergence.

Groundsel (*Senecio vulgaris* L.)

Groundsel is a native annual, ephemeral or over-wintering weed common throughout the UK and found in a range of habitats. It is abundant on horticultural land, where it may occur in vast numbers that can smother a young crop. Plants have a very variable growth habit and leaf shape. Some plants have ligulate ray florets and hybrids occur with Oxford ragwort (*S. squalidus*). Groundsel flowers and sets seed throughout the year, although the main period of flowering is April to October. An individual plant may continue to flower for several months. Flower stems cut down in bud do not ripen viable seed but seeds from plants cut in flower had a germination level of 35 per cent. Groundsel produces around 1,200 seeds per plant.

Fig 48 Gallant soldiers can be easily hoed off when small.

Most seeds can germinate at once and seedlings emerge within a few days of shedding. Freshly shed seeds generally require light but not chilling for germination. However, it has been noted that seeds produced in spring are somewhat more dormant than seeds produced in summer or autumn. Seeds germinate better at lower (10–15°C/50–59°F) rather than higher (20–30°C/68–86°F) temperatures. Seedling emergence generally occurs from March to December, with the main flush from June to October. Field seedlings emerge from the top 3–4cm (1½in) of soil with up to 80 per cent emerging from the surface 5mm (⅕ in). Seedlings are frost tolerant but little germination occurs in winter until the temperature begins to rise. Groundsel can complete its life cycle in five to six weeks but may take longer on nutrient rich soils.

In cultivated soil, 85 per cent of seeds germinated within one year and 100 per cent within five years. In undisturbed soil, groundsel seeds declined by 87 per cent after six years. Losses were due to germination and to seed death in equal amounts. Seeds buried deeper in soil persist longer than seeds in the upper layers of soil. The seeds have a pappus of hairs and are widely dispersed by the wind. The pappus also aids dispersal by adhering to clothes and to animal fur. Groundsel seed has been found in the droppings of various birds and has also been found in cow manure. Groundsel seeds were a contaminant of cereal and vegetable seeds, but not of grass and clover.

Stubble cleaning is effective in dealing with shed groundsel seeds. The surface soil should be cultivated to a depth of 5cm (2in) and the operation repeated at fourteen day intervals. The area around manure heaps, where the weed often occurs in abundance, should be kept clean to prevent groundsel seeds contaminating the manure. Seed numbers in soil may be reduced by around 70 per cent by fallowing for one year. Groundsel

Fig 49 Groundsel seedling.

seedlings with two to six leaves are tolerant of flame weeding. The naturalized rust fungus (*Puccinia lagenophorae*) now occurs widely in the UK and may cause considerable damage to groundsel plants, but there is no guarantee of an attack by this pathogen. Caterpillars of the cinnabar moth (*Tyria jacobaeae*) feed on groundsel in June to July and may weaken or even kill a plant before it can set seed.

Knotgrass (*Polygonum aviculare* L.)

Knotgrass is a procumbent summer annual, frequently troublesome among cereals and root crops. It is often abundant on light, sandy soils and also occurs on roadsides, waste ground and in gardens throughout Britain. It is an aggregate species that can be separated into four closely related but variable species with a similar appearance and commonly recognized as being knotgrass. Flowering occurs from May to October. Average seed production by a large plant may be 6,360 seeds. Seeds are dormant when ripe and for sixty days after maturing; scarification increases seed germination. Dormancy is relieved by low winter temperatures and re-imposed when temperatures rise in late spring. Seedling emergence occurs from late February to June. In the field, most seedlings emerge from the surface 3cm (1in) of soil, with odd seedlings from as deep as 6cm (2½ in).

In studies, 55 per cent of seeds buried in undisturbed soil germinated after twenty years. Seeds in cultivated soil had a mean annual decline rate of 23 per cent and an estimated time of nine to twenty years to decline by 95 per cent. The lifespan of seeds in dry-storage is less than fifteen years. Knotgrass seeds have been a contaminant of cereal, clover and grass seeds,

Fig 50 Knotgrass seedling.

particularly home saved seed. Birds and mammals disperse the seeds, which have been found in cattle, horse and bird droppings and in manure. Seeds have been recovered from irrigation water and were still viable after submergence in water for more than a year.

The plant can regenerate if the top is cut off during active growth. In a comparison of tillage regimes in winter cereals, ploughing and other deep cultivation favoured knotgrass, while shallow cultivations discouraged it. Fallowing for one year has reduced seed numbers in soil by 75 per cent. However, cropping with winter wheat reduced seed numbers to a similar extent, if good weed-control was maintained.

Pale Persicaria (*Persicaria lapathifolia* (L.) Gray)
Pale persicaria is found throughout the UK in waste places and cultivated ground, especially on damp soils. It occurs as a weed in cereals and other arable and horticultural crops on a range of soils. Pale persicaria is very variable in habit, flower colour and pubescence. Hybrids with several related species have been recorded. Pale persicaria flowers from June to October. The seed number per plant ranges from just 10 to 19,300 and demonstrates considerable variation in weight both between and within populations and even between inflorescences on the same plant.

Pale persicaria seeds require a period of after ripening at low temperature before germination will occur. Scarification will promote the germination of after-ripened seeds. The germination of buried seeds is enhanced by exposure to light. The main period of seedling emergence is April to

Fig 51 Pale persicaria plant. One of the bisorts that are difficult to tell apart.

May, with a peak in April. In the field, most seedlings emerge from the surface 4cm (2in) of soil, with the odd seedling emerging from down to 8cm (3in). Few seedlings develop from seeds on the soil surface.

Pale persicaria seeds in the weed seedbank, followed over a six year period of cropping with winter cereals, had a mean annual decline rate of 22 per cent. The estimated time to 95 per cent decline was ten to seventeen years. Seed viability was 50 per cent after four years in dry storage. Seeds submerged in water for five years have retained 59 per cent viability. Pale persicaria seeds have been found as an impurity in cereal and clover seeds. Viable seeds have been found in cattle droppings, although the viability of seeds in manure was reduced to 1 per cent after two months storage. Birds eat the seeds, and seedlings have been raised from the excreta of various species. Seeds have been recovered from irrigation water and may float for up to six months, if the outer covering remains intact.

Redshank (*Persicaria maculosa* Gray)

Redshank is a native summer annual, which is generally distributed throughout the UK on waste, cultivated and open ground. It occurs on most soils but prefers moist, rich soils, acid peaty loams, and is said to thrive on soils deficient in lime. Redshank exhibits considerable variation in leaf shape and size and the flowers may be red, pink or white. It can form hybrids with pale persicaria. Redshank flowers from May to September or until killed by frost. The flowers are usually automatically self-pollinated but can sometimes be pollinated by insects. Seeds ripen from July onwards. The average seed number per plant ranges from 200 to 1,550. The seeds are polymorphic and exhibit considerable variation both in weight and shape.

Seeds are dormant at shedding and for sixty days afterwards. Low temperatures and seed scarification help to break dormancy. Light, nitrate and alternating temperatures interact to promote germination. In the field, seedlings emerge from April to June, with the main peak in April. Field seedlings emerge from the upper 70 mm of soil with most seedlings from the top 40 mm.

Seeds can remain viable in soil for forty-five years. Seeds followed under cropping with winter or spring wheat had a mean annual decline rate of 24 per cent and an estimated time to 95 per cent decline of ten to fourteen years. In dry storage, seeds had 50 per cent viability after three years. Redshank seed has been found as an impurity in cereal, flax, grass and clover seeds. The seeds can pass unharmed through the digestive systems of horses, cattle and deer. The seeds are also ingested and dispersed by birds. Redshank seeds can float in water for twenty-four hours and have been recovered from irrigation water.

Fig 52 Redshank in onions.

In spring wheat, increasing the sowing rate of the cereal and reducing the row width, reduced redshank biomass and seed production. A plant cut back early may persist into a second year. In moist conditions, stem fragments can re-root allowing re-establishment after soil disturbance.

Scented Mayweed (Matricaria recutita L.)
Scented mayweed is a native annual or biennial weed that is locally abundant on sandy or loamy arable soils and waste places throughout England and Wales. Internationally, scented mayweed is grown as a medicinal and industrial crop. Scented mayweed prefers fertile soils that are poor in lime and it can tolerate saline conditions. It is a common weed of cereals and other arable crops, where it benefits from the control of other more competitive weeds. Scented mayweed can be found in flower from May to September but the main flowering period is June to July. It is often the first mayweed in flower. The flowers are insect-pollinated and the time from flowering to seed dispersal is twenty to thirty-five days. The average seed number per plant ranges from 5,000 to 17,000.

Fig 53 Scented mayweed in wheat.

In cultivated soil, seedlings emerge mainly in April to May and August to September; however, odd seedlings can emerge at anytime of the year. Those that emerge from August onwards over-winter as leaf rosettes and flower the following spring. Seeds buried in soil develop a light requirement for germination but this decreases with time. Field seedlings emerge from the top 2cm (1in) of soil with the majority emerging from the surface 1cm (⅖in). Seeds buried in undisturbed soil gave 73 per cent germination after eleven years and dry stored seeds gave 100 per cent germination after three years! Following cereal harvest, scented mayweed seeds are found in large numbers in the chaff. Earthworms ingest the seeds and intact seeds have been recovered in wormcasts. More than 25 per cent of seeds eaten by grazing cattle passed through the digestive system unharmed.

A reduction in seedling emergence has been achieved by cultivating in darkness. Scented mayweed seedlings have shown some tolerance to flame weeding. Scented mayweed is highly attractive to ladybirds and other beneficial insects that feed on aphids.

Scentless Mayweed (*Tripleurospermum inodorum* (L.) Sch. Bip)

Scentless mayweed is a native annual or biennial weed of arable land and waste ground, common on all lowland soils except chalk. It is a frequent weed of cereals, sugar beet and other arable crops. It is moderately resistant to trampling and compaction but it does not thrive at high summer temperatures or in drought conditions. It is intolerant of dense shade and waterlogging. Scentless mayweed is very variable in size and habit, and there is some evidence that winter and summer annual forms occur. Scentless mayweed flowers from June to October. Plants that regenerate in the stubble after cereal harvest may flower again. The flowers are insect-pollinated. The British form is self-incompatable and isolated plants may not set seed. Seeds start to become viable twelve days after flowering and are fully ripe four weeks after the outer florets open. Seed is set from August to October. Each flower head can contain 345 to 533 seeds. A plant may produce 10,000 to 200,000 seeds but figures of over a million have been quoted.

Fig 54 Scentless mayweed in flower.

In the field, scentless mayweed germinates throughout most of the year. The main flush is from February to May, with a smaller one in October. Seedlings that emerge after August over-winter as rosettes and flower in spring. Seedlings and leaf rosettes are frost tolerant. Field seedlings emerge from the top 3cm (2in) of soil with most coming from the surface 0.5cm (⅕ in) layer. Scentless mayweed is shallow germinating because it requires repeated diurnal exposure to light over a period of days to stimulate germination. Scentless mayweed seeds buried in an undisturbed mineral soil gave 52 per cent germination after four years and seeds buried in an undisturbed peat soil for twenty years retained 8 per cent viability. Seeds followed over a five year period of cropping with winter cereals showed an annual decline of 80 per cent. Emerged seedlings represented 15 per cent of the seedbank. Dry stored seeds gave 21 per cent germination after five years.

There is no obvious dispersal mechanism and seed may simply fall to the ground. The seeds may be transported with hay and straw or in mud on tyres and footwear. Scentless mayweed seeds have been a contaminant in grass, clover and cereal seed. Seeds survive passage through ruminants and viable seeds have been recovered from cattle dung. Seeds float for twelve hours in water and retain 75 per cent viability after submergence for 220 days. Cultivation in the dark does not reduce seedling emergence, as the seeds require more than a single light flash to stimulate germination. Seed numbers in soil have been reduced by 50 per cent following a one year fallow.

Pineappleweed (*Matricaria discoidea* DC.)

Pineappleweed was introduced into the UK just prior to 1900 and within twenty-five years it had spread along roadsides throughout most of England. It is now common throughout the UK. It occurs in cereals and broad-leaved arable crops and has become a frequent weed of intensive vegetable crops. Pineappleweed flowers from June to September, sometimes into November. Insects seldom visit the flowers. Seed is set from July onwards. The average seed number per plant ranges from 850 to 7,000. Seed germination is promoted by light and just a short flash is sufficient. Seedlings in the field emerge from February to November, with peaks from March to May and August to October. Plants emerging from January to mid-July set seed and die before winter. Plants that emerge after August over-winter as vegetative rosettes that flower the following spring. Field seedlings emerge from the top 1cm (0.4in) of soil, with the majority emerging from the surface 0.5cm (⅕in). Seeds in undisturbed soil declined by 83 per cent after six years but in cultivated soil the decline was 91 per cent.

Fig 55 Pineapple weed flowers.

Seeds are dispersed in mud and by rain splash. Mud on the tyres of cars was responsible for much of the early spread. The seeds are light enough to be blown by the wind and by passing traffic. Viable seeds have been found in horse droppings. In grassland, pineappleweed is able to colonize areas around gateways and troughs, where livestock have trampled and caused poaching. Seedlings with two to six leaves are tolerant of flame weeding.

Common Poppy (*Papaver rhoeas* L.)

Common poppy is a native annual or over-wintering weed common on arable land, roadsides, waste places and other disturbed habitats throughout the UK. It is a frequent weed of cornfields on light, dry, sandy and gravely soils, and to a lesser extent on heavy land. It occurs mainly in southern and eastern Britain and is rarer in Wales and Scotland. Flowering begins in mid-June, with flushes in late-June and early-July and intermittent flowering then continues to October. Common poppy is normally the last of the poppies to flower. A plant may produce 1 to 400 flowers, depending on the soil fertility and vegetation density. Seeds ripen and are shed three to four weeks after flowering. The mean number of seeds per capsule is 1,360. The average seed number per plant ranges from 10,000 to 60,000 but an isolated plant may have more than 500,000 seeds.

Freshly shed seeds are dormant and require burial in soil for several months to lose dormancy. Seed scarification does not improve germination but light promotes it. Most seedlings emerge from February to April, with a second smaller flush in August to October, although frost may kill the autumn germinated seedlings. Field seedlings emerge from the upper 3cm (1in) soil, with most emerging from the top 1.5cm (⅗in). Seedlings rarely establish in closed communities such as grassland. Soil seedbank numbers of between 2.5 and 20 million seeds per hectare have been recorded for common poppy in vegetable fields in the UK. Seed longevity in soil is more than eight years. In cultivated soil, common poppy seeds had a mean annual decline rate of only 9 per cent and the estimated time to 95 per cent decline was from seventeen to fifty years. Dry stored seeds gave 80 per cent germination after five years. Emerged seedlings represent 8 per cent of the seedbank in any one season.

Common poppy seeds are shaken from ripe capsules by the wind. The seeds travel up to 3m (10ft) initially but are small enough to be further wind-dispersed. Seedlings have been raised from the droppings of birds. Where seeding has occurred, keeping seed on or near the soil surface will encourage germination. Deep cultivations should be avoided. Fallowing has given a gradual reduction in seedbank numbers, but the prolonged seed persistence prevents a rapid decline.

Fig 56 Common poppy in organic cereal.

Long-Headed Poppy (*Papaver dubium* L.)
Long-headed poppy is an annual or over-wintering weed of arable land, especially cornfields and of waste places and roadsides. It has a similar distribution to the common poppy but extends further north and is more frequent in Wales. The distribution in upland areas is limited by the lack of cultivated land but it has been recorded at 427m (1,400ft). The plant can withstand drought and occurs in quite arid areas. Long-headed poppy is very variable and the size, growth habit, capsule number and flower shape of the plant are greatly modified by the environment. Babington's poppy (*P. lecoqii*), in which the sap turns yellow on exposure to the air, is now considered to be a subspecies of the long-headed poppy.

The long-headed poppy flowers from May to July. The flowers are pollinated by bees. Under favourable conditions a large plant may bear 100 to 200 capsules each containing 800 to 900 seeds. The average seed number per plant is given as 5,700 but output can vary from just 10 up to 160,000 seeds. The seeds are inherently dormant and normally will not germinate at all for several months and then germination occurs sporadically over several years. The seeds can remain viable in soil for a very long time. The green capsules on cut plants will produce viable seed. The fully ripe seeds germinate in spring and autumn. The seedlings are susceptible to frost and winter losses can be high.

Shepherd's Purse (*Capsella bursa-pastoris* L.)
Shepherd's purse is a widely distributed native annual (sometimes biennial) weed that grows on most soils, often in large numbers. It is common on cultivated land, waysides and waste places. Leaf shape is very variable and local strains develop due to self-pollination. Flowering occurs throughout the year but is most frequent from May to October. There are ten to twelve seeds per capsule and an average of 4,500 seeds per plant, although much higher numbers have been recorded. Seed size varies considerably both within and between populations. Plants growing in adverse conditions produce fewer but larger seeds. Flower spikes cut prematurely produce viable seeds from the large, but not from the small, immature, fruits. Seeds require a period of after ripening before they will germinate. In cultivated soil, seedlings emerge from February to November, with peaks of emergence in May and September. Some plants produce seed within six weeks of emergence and three generations can occur in a year. Field seedlings emerge from the top 2cm (1in) of soil with 89 per cent from the surface 1cm (⅜ in). Seed buried in undisturbed soil can remain viable for thirty-five years or longer.

Seeds are passively dispersed around the parent as the fruits split open but are small enough to be blown further by the wind. Long-distance

Fig 57 Shepherd's purse under organic cereal.

transport is facilitated by the thin layer of sticky mucilage that forms when the seeds are moistened. The seeds may adhere to animals and can be carried in mud on implements, boots and tyres. Shepherd's purse seeds were a common contaminant of clover and grass seed. Viable seeds have been found in manure and in the droppings of birds, cattle and goats, and seeds ingested by earthworms have been recovered unharmed in wormcasts on the soil surface. Seeds have been found in irrigation water and can remain viable after submergence for over nine months.

Stubble cleaning can be an effective way of reducing seed numbers in soil. The surface soil is cultivated to a depth of 5cm (2in) at fourteen day intervals. Fallowing has little effect on seed numbers in soil in the short or the long-term. In reduced tillage systems, seed numbers increase in the upper 15cm (6in) of soil. Seedlings with two to six leaves are tolerant of flame weeding. The seeds are susceptible to soil solarization. Geese will eat shepherd's purse and may control it selectively in certain crops. The mature plant is attacked by several fungal diseases and the seedlings are prone to flea beetle damage at the cotyledon stage.

Small Nettle (*Urtica urens* L.)

Small nettle is found on cultivated ground and waste places, particularly on light soils. It is troublesome on horticultural land and causes problems in crops that are hand harvested. It prefers soils with a high organic matter content and is intolerant of heavy shading. Small nettle flowers from June to November and the flowers are wind-pollinated. Seed is set from June onwards but plants continue to grow and flower until killed by frost. The average seed number per plant is around 1,000 but a large plant can have up to 40,000 seeds. Buried seeds require light for germination but just a five second flash is sufficient. Seeds germinate well in partial shade but bright light inhibits germination. Small nettle seedlings emerge from March to October, with peaks in April and July. Most seedlings emerge from the surface 2–3cm (1in) with the odd seedling emerging from a depth of 6cm (2½ in). Small nettle seeds in undisturbed soil declined by 61 per cent after six years but in cultivated soil the loss was 96 per cent. Viable seeds have been found in cattle droppings.

Small nettle is absent from habitats that are mown or grazed. Seedling numbers may increase following an application of manure. Small nettle seedlings with two to six leaves are susceptible to flame weeding. The seeds are killed by soil solarization.

Prickly Sow-Thistle (*Sonchus asper* L.)

Prickly sow-thistle is a troublesome annual or over-wintering weed, common on arable land and other disturbed habitats. It grows on most soils but prefers well drained slightly acid to alkaline soils, and has some tolerance of saline conditions. It is generally less abundant than the smooth sow-thistle. Prickly sow-thistle flowers from June to October. The flowers are self-compatible and mature seeds are formed one week after flowering. The average number of seeds per flower head is 198 and a plant can have a 100 flower heads. Seed numbers per plant range from 21,500 to 25,000 but a large plant may have 60,000. Seeds on the soil surface germinate better than those buried at 3cm (1in) deep. Seedling emergence occurs from March to November, with peaks in March to April and August to November; but a few seedlings can germinate at any time. The half-life of seeds in cultivated soil is just one year and in dry storage it is two to three years.

The seeds are normally wind dispersed but under damp conditions the hairy pappus collapses and dispersal is prevented. The seeds are eaten by birds and viable seeds may be found in their droppings. Viable seeds have also been found in cow manure, and seeds ingested by earthworms have been recovered intact in the worm casts. Prickly sow-thistle seeds have been a contaminant in clover, grass and cereal seeds.

In grass, prickly sow-thistle may be controlled by grazing with sheep or by mowing. Plants that are cut down early in the year can produce further flower stalks. The seeds are susceptible to soil solarization.

Fig 58 Prickly sow-thistle against a rabbit fence.

Smooth Sow-Thistle (*Sonchus oleraceus* L.)

Smooth sow-thistle is a native annual or over-wintering weed, common throughout the UK and, like the prickly sow-thistle, it is a pioneer species that colonizes disturbed ground. It exhibits considerable variation in leaf form and flower colour. Smooth sow-thistle flowers mainly from June to August but flowering can continue until the first frosts. The flowers are self-fertile. Mature seeds are formed one week after flowering. Plants cut in bud do not ripen viable seed but the seeds from plants cut in flower may be 100 per cent viable. The average seed number per flower head is 140 and the number of flower heads per plant is around 44. The potential seed number per plant varies considerably with environmental conditions and estimates range from 5,000 to 40,000. Seedlings emerge from April to September but the main flush occurs from April to June. Winter rosettes can withstand moderate frosts.

The half-life of seeds in cultivated soil is short and may be just one year. Dry stored seeds remain viable for around ten years. The seeds have a pappus of hairs and are wind dispersed in dry conditions. Smooth sow-thistle seeds can occur as a contaminant in clover, grass and cereal seeds.

Fig 59 Flower of smooth sow-thistle.

The seeds form an important part of the diet of native birds, and seedlings can emerge from the droppings.

Topping plants to prevent seeding may result in the production of further flower stems. Smooth sow-thistle is favoured by minimum tillage systems. It cannot withstand repeated trampling and may be controlled by sheep grazing or by mowing. Smooth sow-thistle does not survive beyond the seedling stage if it is shaded to any great extent.

Common Field Speedwell (*Veronica persica* Poiret)
Common field speedwell was introduced into the UK with clover and other crop seeds, and was first recorded in 1825. It has become the commonest speedwell and one of the most common annual weeds in the UK. Field speedwell is found on all types of cultivated soils but prefers nutrient-rich loams of pH 6.0 to 8.0. It flowers throughout the year and is self-fertile. The flowers are visited by insects but are often self-pollinated. Seeds are set mainly from June to October but plants can be found in fruit at any time of year. The average number of seeds per plant is 2,000 but a large plant may produce up to 7,000 seeds. There can be two generations in a year.

In the field, seedlings emerge from February to November, with peaks of emergence in May and September. The majority of seedlings emerge from the surface 3cm (1in) of soil with the occasional seedling emerging

Fig 60 Common field speedwell.

from down to 6cm (2½ in). The seedlings are frost tolerant. The annual seed decline in a series of autumn sown crops was 46 per cent and the time to 99 per cent decline was calculated at six years. In studies, the annual seedling emergence represented 4 per cent of the seedbank. Under a grass sward, the mean annual seed decline was 18 per cent and the half-life was three years six months.

There is no obvious seed dispersal mechanism. Seeds are dispersed as a contaminant in crop seed, manure and fodder. Earthworms ingest common speedwell seeds and viable seeds have been recovered from worm-cast soil. Stem fragments will root readily in cool, moist conditions. Growth of the weed is strongly suppressed in shade. A one-year fallow did not reduce seed numbers in soil.

Ivy-Leaved Speedwell (*Veronica hederifolia* L.)

Ivy-leaved speedwell is common in cultivated ground throughout the UK. It is frequent in winter cereals and two subspecies occur: one with smaller lilac flowers and the other with larger blue flowers. The former occurs mainly in gardens and shady places, the latter in cultivated fields and by roadsides. Ivy-leaved speedwell flowers from March to June, sometimes into August. The average seed number per plant ranges from 40 to 400. Ripe seeds are dormant but after one to two months burial in soil, they will germinate at low temperatures. Dormancy is overcome by warm summer temperatures and gradually re-imposed by low winter temperatures. The main period of seedling emergence begins in October and continues until May. Seedlings emerge from the top 11cm (4in) of soil with most emerging from between 0.5 and 6cm (⅕–2½in) deep. Few seedlings emerge from the surface 0–0.5cm (0–⅕ in).

Seed longevity in soil is three to four years and seeds in cultivated soil had an annual decline rate of 57 per cent. The estimated time to 99 per cent decline was five years five months. Emerged seedlings represented 2 per cent of the seedbank. In grassland soil, ivy-leaved speedwell seeds had a mean annual decline of 19 per cent and a half-life of three years six months. Viable seeds have been found in cattle droppings. The weed has been found to decline following a series of spring cereals. Fallowing for a year generally reduces weed seed numbers in soil but there is a similar reduction in seed numbers following a well-managed cereal crop.

Wall Speedwell (*Veronica arvensis* L.)

Wall speedwell is found on walls, banks, open ground and cultivated land. Wall speedwell prefers nutrient rich, moderately acid, sandy loams in the pH range 6.0 to 8.0. It is a plentiful weed in autumn sown wheat. Wall speedwell flowers from March to October. Seed is set from June onwards. Wall speedwell has around eighteen seeds per seed capsule and

Fig 61 Ivy leaved speedwell.

there are 17,000 seeds on a large plant. Fresh seed is generally dormant. Seedlings emerge in two main flushes, April to May and August to November, but odd seedlings emerge at other times too. The optimum depth of emergence is 0–0.5cm (0–⅕ in) and the maximum is generally 1cm (⅖ in). Seedlings are frost tolerant. Most seeds germinate in the first three years after shedding and few viable seeds remain after five years in cultivated soil. The annual percent decline is around 58 per cent. The seeds are small enough to be blown around by the wind. Viable seeds have been found in cattle droppings.

Wall speedwell does not decline following a change to minimum cultivation. Seed numbers in soil were reduced by 50 per cent after a one-year fallow. The seeds of wall speedwell are consumed by several species of ground beetle.

Wild Radish (*Raphanus raphanistrum* L.)

Wild radish is a troublesome annual or biennial weed that occurs on cultivated and rough ground. It is common throughout the UK and prefers a nutrient rich, lime free, sandy or loamy soil. Yellow and white flowered forms are found, some with violet veining of the petals. The different flower colour forms often occur in a single population. Wild radish flowers from May to October. Seed set is often variable due to self-incompatibility and erratic pollination. Seeds within a pod may be the result of fertilization with pollen from different parent plants, so genetic differences may be present even in a single pod. Each pod contains from one to ten seeds. The seedpods are indehiscent and break up into segments containing a single seed. Seed numbers per plant range from 160 to 1,875.

Freshly shed seeds do not germinate readily. The outer pod tissue inhibits germination both mechanically and chemically. Seeds of the white and purple veined flower forms are more dormant than the yellow ones. Light has little effect on germination. Cultivation that leads to shallow burial in moist soil, stimulates germination. Seedling emergence occurs from March to October, with the main flush in March to May and a smaller one in August. In the field, seedlings emerge from the top 8cm (3in) of soil, with the majority from the surface 2cm (⅘ in)

After two years burial in soil, seed viability was 53 per cent at 10cm (4in) depth and 16 per cent at 1cm (⅖ in). Under a grass sward, the seeds had a mean annual decline rate of 33 per cent and a half-life of two years. Wild radish seeds have occurred as an impurity in cereal seeds, particularly home saved seeds. The seeds and seedpods release a toxic vapour during storage that reduces the viability of crop seeds kept in close proximity. Seeds have been found in cattle droppings and manure, and seedlings have been raised from the droppings of various birds.

Fig 62 White-flowered wild radish.

Wild radish has a long taproot that aids drought-resistance and provides adequate reserves for regrowth if the plant is cut down. Wild radish seedlings are attacked by flea beetles and can suffer the same damage as cultivated brassicas. Stubble cleaning is an effective way of dealing with freshly shed wild radish seeds. Soil should be cultivated to 5cm (2in) depth at fourteen day intervals. Fallowing is generally unsuccessful because wild radish seeds have different levels of dormancy.

Annual Grass Weeds

Grass weeds, like broad-leaved weeds, compete their life cycle in one season or year. They generally have less persistent seeds than the broad-leaved weeds but are favoured by minimum cultivation systems. The basal growing point gives grass seedlings protection against flame weeding. Annual grass weeds suffer from ergot and this can result in contamination of the cereal grain.

Annual Meadow Grass (*Poa annua* L.)
Annual meadow grass is common throughout the UK. It is one of the main weeds on cultivated soils and in grassland. It is an important constituent in the diet of many farmland birds. Although the plant is relatively small, it often emerges in sufficient quantities to smother crop seedlings. Annual meadow grass is shade tolerant and can withstand considerable trampling but it is sensitive to drought. Flowering and fruiting

can occur at any time of year. Over-wintered plants start to flower in May, while seedlings that emerged in spring flower from July to September. Seed is produced abundantly from April to September, although high temperatures can limit seed set. Seed number per plant is said to average 2,050. Annual meadow-grass can flower and set seed in short turf, even when mown regularly.

Annual meadow grass seedlings emerge from February to November, with the main peaks of emergence in early spring and in autumn. Seedlings that emerge from August to December over-winter and begin to tiller when growth resumes in spring. Annual meadow grass seeds can remain viable in soil for at least four years but decline faster in cultivated soil. Dry stored seeds were still fully viable after three years.

Annual meadow grass is a prolific seed producer but there is no obvious dispersal mechanism. Most seeds fall around the parent and soon become incorporated into the soil. Mowing when the weed is in flower will also spread the seeds. Viable seeds have been found in horse and cattle droppings, but viability is lost in manure after a period of storage. Annual meadow grass seed has been a contaminant in cereal and cultivated grass seed.

Fig 63 Annual meadow grass.

Young seedlings have fibrous roots and are easily dislodged by cultivation. In winter cereals, annual meadow grass populations increase where direct drilling is the practised. However, ploughing leads to greater seed persistence. Repeated mowing and close grazing of annual meadow grass merely causes increased branching and does not prevent seed set. Mature plants are able to survive uprooting and may continue to ripen seeds.

Barren Brome (*Anisantha sterilis* (L.) Nevski)
Barren brome is common throughout the UK. It is native in rough and waste ground, in hedgerows and by roadsides. It can spread from the hedge bottoms and field margins into the arable field, where it may form dense patches. It has increased as a weed of cultivated land since the 1970s, being favoured by continuous cropping with winter cereals and the adoption of non-plough techniques. Barren brome flowers from May to July. Seeds mature rapidly and are shed from late June to early August. Over 200 seeds are produced per plant. There may be some initial dormancy but the majority of seeds will germinate immediately on contact with a moist soil, especially if buried. Dry conditions and sunlight inhibit germination and encourage the persistence of seeds left on the soil surface. Autumn rains usually promote a flush of germination. If drought or low temperatures prevent germination in the autumn, seeds may persist and germinate the following spring. Seedlings require a period of cold to vernalize them in order to flower. Spring emerging seedlings that do not become vernalized, may fail to flower before crop harvest. Most seedlings emerge from the top 5cm of soil.

The majority of seeds do not persist in soil for longer than twelve months. However, seeds that have become dormant after remaining on the soil surface, may survive for a longer period. Some seeds stay attached to the flower stem until January; these seeds are slower to germinate and may represent an additional survival strategy for the plant. At cereal harvest, barren brome seeds are widely dispersed during combine harvesting. The seeds can be an impurity in crop seeds, particularly home saved winter cereal seeds. Viable seeds have been found in horse droppings.

Where barren brome is in the headland, care should be taken not to allow it to spread into the field. The inclusion of spring sown crops can help to break the cycle and reduce infestations. In a moist autumn, there may be an opportunity to kill barren brome seedlings, if the drilling of winter cereals is delayed. The burial of freshly shed seeds avoids dormancy developing due to light exposure. Seedlings are unable to emerge from below 13cm (5in) so ploughing can be very effective, if inversion is complete. Minimal tillage will encourage barren brome to increase in numbers.

Fig 64 Barren brome in headland.

Black Grass (*Alopecurus myosuroides* Huds.)
Black grass is rarely found outside of cultivated land and is most abundant in winter crops on heavy land. It flowers from May to August, sometimes later. Seed production ranges from just 50 to over 6,000 seeds per plant. Seeds ripen quickly and many are shed before crop harvest. Some seeds are able to germinate soon after shedding; others remain dormant for a few months. Shallow burial can result in some degree of dormancy. Light and fluctuating temperatures appear to increase seed germination. If conditions are favourable most seeds germinate in late October and early November. There is a small flush of seedlings in spring, which is greater if autumn germination has been prevented by soil conditions. Small seedlings can tolerate moderate frost and those with tillers can withstand freezing to –25°C (–13°F). Seedlings also survive waterlogging in the autumn.

Seed decline is greatest in the first year but appreciable numbers of seeds can remain after four years. If seeds are ploughed down deeply they may retain viability for eleven years. Seeds in dry storage will still germinate after thirteen years. Seeds have been found as a contaminant in

Fig 65 Black-grass, in this case in brassicas, where it is easy to control.

cereal, grass and clover seed samples. Seeds do not usually survive passage through birds or animals, but they have been found in cattle droppings.

Early drilling of winter cereals leads to severe infestations because it coincides with peak black grass emergence and the weed has time to become fully tillered before winter. Sowing cereals before 25 October has been shown to increase black grass infestations; sowing after 5 November has led to a decrease. However, there can be a substantial loss in crop yield if winter wheat is drilled after mid-November. Where there is only a small reserve of seeds in the soil, ploughing generally reduces the level of infestation from fresh seeds, while tine cultivations leave most seeds ready to germinate in the surface layers of soil. However, a field that has suffered black grass infestations for many years will have seeds present throughout the soil profile and ploughing will unearth the buried seeds in large numbers. It has been suggested that, to prevent a likely build-up of black-grass under minimum tillage systems, the land should be ploughed every five years. Fallowing can reduce seed numbers in soil dramatically and an autumn fallow is ideal for this. However, seed numbers will build up again if further seed-shedding occurs. Observations tend to indicate that black grass decreases in organic rotations and may not be as troublesome as in conventional cereal production.

Wild Oat (*Avena fatua* L.)

Wild oat was probably introduced into Britain as a contaminant of cereal seeds and is now common throughout much of the country. It occurs on most soils but is particularly troublesome in cereals on heavy land. Wild oat flowers from June to October. The first wild oat panicles become visible above the cereal crop in early June but flowering and panicle production can continue up until harvest. Seeds become viable around ten days after fertilization. In the absence of competition, a single well-tillered plant can produce up to 2,000 seeds. However, in a cereal crop, seed numbers of 60 to 200 per plant are more likely. Wild oat seeds are shed as they ripen and this occurs over an extended period. The later a cereal crop is harvested, the fewer the wild oat seeds that remain to contaminate the grain.

Fresh seeds are dormant, or rapidly become so, but the level of dormancy varies between populations. Seeds from the smaller secondary floret are usually more dormant than the primary seeds. In the autumn, the level of dormancy declines and is induced again in late spring. Damage to

Fig 66 Wild-oats rising above organic wheat.

the seed coat can relieve dormancy at any time by allowing oxygen to reach the seed. Wild oat has two main periods of germination: a small autumn flush from September until early-November, and a main flush from January to early-May. Most seedlings originate from seeds in the top 8–10cm (3–4in) of soil but some emerge from a depth of 15–24cm (6–10in). Seedlings are frost sensitive until they reach the three-leaf stage.

When seeds are left on the soil surface, viability declines within a few months due to germination, predation and fungal attack. Freshly shed seeds incorporated into the soil can remain dormant but viable for up to six years. If soil is cultivated regularly, the majority of seeds will only survive for two to three years. Wild oat seed can be dispersed by farm machinery and continues to be spread as a contaminant of cereal seed and straw. The seed does not generally survive in silage or in the digestive systems of cattle.

Wild oat control is improved by shallow cultivations in autumn to induce a proportion of seed to germinate and by deeper working in spring to kill the emerged seedlings. If a large number of seeds remain on the soil surface after cereal harvest, early ploughing will lead to more seeds surviving than if cultivations are delayed. Preparing the seedbed well in advance of drilling will allow the earliest wild oat seedlings to be killed by cultivation. Hand rogueing of cereals is possible with weed populations of 400 to 500 wild oat plants per hectare (160–200 plants per acre). A summer fallow will have little effect on wild oat seeds in soil, as germination is unlikely to occur at high temperatures. For the same reason, crops sown after May have few wild oats in them.

BIENNIAL WEEDS

Biennials generally have a life cycle that stretches over two years or growing seasons. They grow vegetatively in the first year, over-winter and usually flower and set seed and die in the second year. In unfavourable conditions, or following plant injury, flowering may be delayed for one or more years. For this reason, biennials are sometimes referred to as monocarpic perennials. If flowering is prevented repeatedly, a plant may bulk up sufficiently to become a short-term perennial.

Managing Biennial Weeds

In annual crops, biennial weeds are unlikely to set seed but can grow large enough to cause yield losses. In arable crops, the seedlings will be destroyed by the surface cultivations used to control annual weeds. In grassland, biennial weeds require the same types of controls as those that are used to deal with perennial weeds.

Common Biennial Weeds

Common Ragwort (*Senecio jacobaea* L.)
Common ragwort is a weed of wasteland and pasture that occurs in every county in the UK. Ragwort is rarely a problem in cultivated fields and is not found on acid, peaty soils. Marked fluctuations can occur in ragwort populations, and plant numbers may suddenly increase or decrease for no apparent reason. The flowering period is long, starting in mid-June and continuing until November. When allowed to flower and set seed, ragwort is usually a biennial, but if it is defoliated, it behaves as a perennial. Ragwort will only flower after exposure to winter cold but rosette size is also important as plants must attain a threshold size and this may take more than three years. The flowers are insect-pollinated. The average seed number per flower head is seventy but the number of flower heads per plant can range from less than 200 to over 2,000 and a single plant can produce 150,000 seeds. The seeds may begin to ripen in July/August but are shed chiefly from September onwards. Viable seeds can be formed on cut-down flower stems, if the flowers are open when severed.

Two types of seed are produced in a flower head, each with different germination characteristics. Seeds from the disc florets at the centre of the flower head are lighter and have a pappus of hairs to aid long-distance dispersal to new sites. The disc florets around the edge of the flower head produce heavier seeds with thicker coats and no pappus. These remain *in situ* until shaken free and fall close to the parent plant. They remain dormant in the soil ready to germinate when the parent plant dies and leaves a gap in the vegetation. Seeds left on the soil surface germinate mainly in autumn but a few seedlings also emerge in spring. In cultivated soil, seedlings emerge from February to September, with a main flush in April and a lesser flush in August to September.

Seeds buried in soil can remain dormant for over sixteen years; those near the surface or in cultivated soil persist for less than four years. Birds may eat the seeds but viable seeds are rarely found in bird droppings. Seeds that are eaten by sheep pass through the digestive system undamaged. In dry conditions, the seeds from the disc florets are carried by the wind. Ragwort seeds may be dispersed by water. Initially the seeds float, then sink but float again as they begin to germinate. Ragwort seeds can be dispersed in hay and have occurred as a contaminant in crop seeds.

An established ragwort plant growing in pasture has a spreading rootstock and often forms a group of offsets. The taproot is replaced by a system of adventitious roots that acts as a source of new growth following plant damage to the parent plant. These roots, or small fragments of them, produce adventitious shoots that colonize adjacent spaces. Severed but undisturbed roots are more likely to regenerate than disturbed ones. Severe frost may kill the above ground plant parts but regeneration usually occurs from the crown. Ragwort does not tolerate flooding.

Toxic alkaloids are present in all parts of the plant and it regularly causes loss of livestock through liver damage. Cattle and horses usually avoid ragwort when there is adequate grazing but newly turned out stock may eat it if hungry. Sheep are partial to it in the young state and appear to be more resistant to the poison than cattle, but they are not immune. The presence of ragwort in hay, silage or dried grass is the main source of poisoning. Drying or similar processes do not affect the poisonous properties. Wilted plant material is more palatable to stock than the growing plant but it is as toxic. Hoary ragwort (*S. erucifolius*) and marsh ragwort (*S. aquaticus*) also contain poisonous alkaloids. In the UK, the Ragwort Control Act 2003 has led to the provision of a code of practice, published by Defra, that gives guidance on preventing the spread of ragwort in

Fig 67 Well-developed rosette of ragwort.

situations where it is likely to be a danger to horses and other livestock. The code does not seek to eradicate ragwort, but only to control it where there is a threat to the health and welfare of animals. Guidance on the disposal options for common ragwort has been prepared to supplement the advice given in the Code of Practice and should be read in conjunction with it.

In grassland, maintaining a dense vigorous sward on well drained land will prevent ragwort becoming established. Mowing is at best a short-term measure to prevent seed production as, in reality, it causes plants to perennate and persist. Pulling is more effective but is only feasible for small infestations and regrowth can occur from detached roots left in the soil. Mechanical pulling with the 'eco-puller' is possible when the flower stem has elongated and there is sufficient height difference with the crop. Pulled material must be disposed of safely. Controlled grazing to maintain optimum sward height can help to keep the weed in check. Plants are weakened by winter/spring grazing by sheep but there is some risk to the animals on heavily infested fields. Seed set, in particular, is reduced because sheep graze the young rosettes and prevent flowering. Sheep grazing should be practised only at the rosette stage, with a low ragwort infestation and as long as other herbage is also available. Poultry have been seen to scratch out and eat the ragwort rosettes down to the roots but there is rapid regeneration. Ploughing and the introduction of an arable rotation is the most effective method of control in heavily infested pasture. Grass may be resown after a period of annual cropping but direct reseeding will often result in a rapid re-infestation. Reseeding is best carried out under a cover crop. The presence of clover and adequate phosphate levels are said to discourage ragwort establishment.

More than 200 species of insect and other invertebrates have been found on common ragwort in the UK. The flowers are among the most frequently visited by butterflies in Britain. Caterpillars of the cinnabar moth (*Tyria jacobaeae*) feed on ragwort in June to August and may weaken or even kill a plant before it can seed. The caterpillars are likely to have the greatest effect on ragwort when summer rainfall is low and the regenerative ability of defoliated plants is reduced by moisture stress. The root-feeding flea beetle (*Longitarus jacobaeae*) can cause severe damage to ragwort. The larvae attack the root crown and feed externally on lateral roots. Various rust fungi and other pathogens also infect ragwort but do not cause serious injury.

Spear Thistle (*Cirsium vulgare* (Savi) Ten.)

Spear thistle can be a serious problem in grassland and waste and cultivated ground throughout the UK. It is absent from densely shaded and waterlogged habitats. Seedlings and young rosettes are relatively susceptible to drought. Spear thistle flowers in June and July but rosettes can remain without flowering for up to four years. The first ripe seeds are formed by the end of July. Plants cut down in flower produce a few seeds but none are viable. The average number of seeds per flower head is around 100 but there can be up to 340. Seed production per plant may vary from 1,600 to 8,400 seeds. Seeds are dispersed during August and September.

Seeds have little dormancy and germinate rapidly in moist conditions, at favourable temperatures, in the light. Spear thistle seedlings emerge in spring and autumn but the main period of emergence is March to April. In pasture, seedlings emerge from small areas of disturbance or from the bare patch left when a parent plant dies. Once a leaf rosette develops, it physically suppresses the growth of the surrounding vegetation. The rosettes grow better in grazed than ungrazed pasture because of reduced competition from neighbouring plants. Spear thistle is propagated only by seeds. The seeds persist for less than a year on the soil surface; however, buried seeds survive for at least three years. Seeds in dry storage remain viable

Fig 68 Spear thistle in grassland.

for up to three years. The seed has a feathery pappus to aid wind dispersal but most seeds are dispersed less than 2m (6½ ft) from the parent. Seed is also spread as a contaminant in fodder and in crop seed.

In grassland, young plants should be dug out at the rosette stage or cut down to prevent seeding. The taproot may be cut using a thistle hoe or plants may be pulled mechanically using the 'eco-puller' set low to the ground. In permanent pasture, mowing just as flowering begins may kill spear thistle. Sward management should aim to encourage a dense cover of grass to prevent spear thistle seedlings becoming established. Spear thistle is unpalatable to most stock but some sheep breeds will graze young rosettes. Dorset horn sheep have been seen to eat it and dig for the roots. Prior to dispersal, seeds within the flower head are predated by insect larvae. Various finches have been seen taking seeds from the flower heads. After dispersal most seed predation is by mice and voles; birds and insects are less important. Under the Weeds Act 1959, Defra has the powers to require an occupier of land to prevent the spread of spear thistle.

PERENNIAL WEEDS

Perennial weeds include both broad-leaved and grass types. Perennial weeds are thought to increase in organic systems over time. Whether this is the expansion of existing infestations or the onset of new weed problems, and whether it is a result of the changes in weed management that follow the conversion to organic growing methods, is not clear.

Perennial weeds have developed a variety of strategies that enable them to persist, multiply and spread. The different species rely to a greater or lesser extent on both seed and vegetative means to enable them to succeed in relatively stable and disturbed habitats. Some perennial weeds are a problem in both situations, while others cannot cope with disturbance or cannot compete with an established stand of vegetation. Stationary perennials, which generally multiply by seed and have stout perennating taproots, tend to be more problematic in grassland or other more permanent crops. In contrast, creeping perennials, which tend to extend across the ground by means of horizontal shoots (rhizomes, stolons or thickened roots), can be difficult in either situation but especially in arable systems, as fragmentation can rapidly increase the number of plants. The regenerative capacity of the vegetative parts is also an important feature of the survival strategy of perennial weeds. Seed production is more important in some species than others but its significance should not be discounted.

Managing Perennial Weeds

The location of vegetative organs is an important factor to take into consideration when contemplating weed management by mechanical means. Underground organs may be located at a considerable depth or may be within the plough layer. Plant morphology will also determine how competitive the weeds are in any cropping situation.

The principal methods of non-chemical control for perennial weeds are cultivation, cutting and competition. No single method of control, and still less a single weeding operation, will give adequate control. A sustained approach is needed for perennial weed management. The aim is to prevent seeding and to starve out the perennial organs; this is often accomplished, in the first instance, by repeated mowing or cutting. Historically, fallow season tillage has been used to control severe infestations of perennial weeds. Like repeated mowing, it starves the underground organs and prevents seed production. Biological control has obvious appeal and has received much attention but it is unlikely to provide reliable control. However, natural control, such as seed predation that goes on in the background, is important in supporting other control measures.

Broad-Leaved Perennial Weeds

Broad-leaved perennial weeds contain species with both stationary and creeping habits.

Creeping Buttercup (*Ranunculus repens* L.)
Creeping buttercup is the commonest of the buttercups and occurs in damp meadows and pastures throughout the UK. It forms large colonies on wet, heavy land. In a ridge and furrow pasture it often occurs in a band along the bottom of the furrows. It can withstand trampling and compaction, and is often found in gateways and on paths. It can tolerate both waterlogging and a moderate drought. Some plants spread widely, while others remain relatively compact. Although it is usually avoided by stock, creeping buttercup is more palatable than the other buttercups and may be grazed. However, it can cause diarrhoea in sheep and cattle.

Creeping buttercup flowers from May to August. The flowers are insect-pollinated and cross-pollination predominates. However, not all plants flower, and those that do may have only a few flowers each producing twenty to thirty seeds. A flowering shoot may yield 140 seeds and the average seed number per plant is 687. Plants from ruderal habitats tend to flower more freely, perhaps because seed production is more beneficial in a disturbed situation. Ripe seeds are highly dormant and may need a period of after ripening. Seedlings emerge sporadically throughout the year but the main periods of emergence are from February to June and August to October. Most seedlings emerge from 0.5 to 3cm (⅕–1in) deep in soil. Seedlings are unaffected by frost.

Creeping buttercup has a short swollen stem base, long stout adventitious roots and strong leafy, epigeal stolons that root at the nodes. The stolons begin to develop around the time of flowering. In open and fertile sites, the stolons are long and well branched to ensure rapid colonization. In close turf the stolons are few. It has a creeping habit under intense grazing or mowing but makes erect growth in taller vegetation. Daughter plants form in the axils of the stolon leaves. The stolons wither away leaving the daughter plants as independent units. The parent plant dies after seed ripening and the daughters over-winter as leaf rosettes.

Seed buried in undisturbed soil remains viable for over twenty years. There is considerable persistence of creeping buttercup seed in the soil under grassland. Seeds have survived three years in dry storage. Seedlings have been raised from the excreta of various birds; they have been found in the droppings of cattle and horses; and seeds eaten by

Fig 69 Creeping buttercup.

earthworms have been recovered from wormcasts. Creeping buttercup seed has been a contaminant in clover, grass and cereal seeds. In dry conditions, creeping buttercup sets seeds but in wet conditions it relies on vegetative reproduction for spread and persistence.

Creeping buttercup is controlled by frequent and vigorous cultivation in hot weather. Plants can survive shallow burial. Sward destruction in spring provides ideal conditions for rapid colonization by creeping buttercup seedlings. After ploughing, it is best to clean up the land with one or more root crops before putting it down to grass again. In grassland, small patches can be removed manually. Meadows should be harrowed in spring to drag out and gather up the runners. Intense grazing prevents seed set.

Bulbous Buttercup (*Ranunculus bulbosus* L.)

Bulbous buttercup is found in dry grassland and on grassy slopes. It is primarily a lowland buttercup and is found on the ridges in ridge and furrow pastures. Seedlings fail to establish in very wet conditions. Bulbous buttercup has an annually renewed swollen stem base from which arise one or more aerial shoots. It can readjust its depth by up to 10cm (4in) if covered with loose soil. It is often abundant on well drained or even thin soils, due to the ability to survive summer drought. Bulbous buttercup is the most toxic of the buttercups to livestock. The foliage is harmful if eaten fresh but is not poisonous when dried in hay. Cattle and horses usually avoid the plant but poisoning can occur with new stock that has not come across it before. The seeds and stem base form a major part of the diet of partridge.

The first flowers appear in March, but peak flowering is normally from mid-May to June. Each flower head contains twenty to thirty seeds. The mean number of seeds per plant is around seventy. Often seed will have ripened and been shed by early July. Fresh seed is normally dormant and requires a period of after-ripening. Seedlings emerge from July to October but most emerge in the autumn. The optimum depth of emergence is 1.6cm (¾ in). After seed has set in July, bulbous buttercup dies down leaving a bare patch in the grass. It passes the summer as a 'corm' that may remain dormant from July to September. Corms do not need a period of cold to break dormancy, only adequate moisture and favourable temperatures. The lateral bud of a corm that has flowered, or the terminal bud of a corm that has not, becomes active and grows out from the old corm and extends up to the soil surface. Leaves develop and a small rosette is formed that over-winters. The vertical corm like stock over-winters and the rosette produces new leaves in March. A new corm develops at the base of the shoot. Food reserves are transferred to the new corm in February/March and the old corm dies.

Plants do not spread vegetatively unless the new shoot is damaged and auxiliary shoots develop. The plant can also re-establish itself after ploughing. Seed is the main method of reproduction. There is no obvious dispersal mechanism and seeds normally fall around the parent. Cattle may eat the seed heads during grazing and the seeds can pass unharmed through the digestive system. The seeds also survive digestion by birds.

Bulbous buttercup becomes established where overgrazing has occurred or bare patches develop due to urine burn or dung heaps, and where fresh soil is exposed due to disturbance. The plant cannot tolerate early competition from taller plants and bulbous buttercup seldom persists in grass cut for hay or silage. Bulbous buttercup fails to survive in soil that remains wet for extended periods. It is intolerant of trampling and is usually absent from paths that traverse grassland. Pigs eat the corms with relish and do not appear to be harmed. Geese pull the plants up. Other birds eat the plant and seeds, and there is predation too by voles and mice.

Meadow Buttercup (*Ranunculus acris* L.)

Meadow buttercup is native in grassland especially on damp soils. It is recorded up to 1200m (4,000 ft) in Britain. It is characteristically found on grazed and mown grassland. Meadow buttercup is a serious weed of old permanent grassland and its abundance is considered an index of the age of the pasture. It prefers intermediate conditions of moisture and drainage, and is found on the slope of the ridges on ridge and furrow grassland. It is not tolerant of trampling but can survive frequent cutting. Meadow buttercup is a very variable species with several sub-species and varieties but only ssp. *acris* is native in Britain. Cattle usually avoid the plant but meadow buttercup can cause inflammation of the digestive system if eaten fresh but not when dried in hay.

Meadow buttercup flowers from May to July and sometimes longer. Some populations are self-sterile and insects visit the flowers. The first fruits appear in mid-June and peak fruiting is in mid-August. Each flower head contains thirty seeds and seed numbers per plant normally range from 200 to 1,000 but a large plant may have 22,000 seeds. Germination is said to occur in autumn and spring but most seedling emergence is from January to April.

Meadow buttercup passes the winter as a rosette of small leaves that appear unaffected by frost. The plant has a short, creeping rhizome. In the autumn, a small branch of the rhizome produces a new vegetative shoot close to the parent. The connection decays as the plant develops, resulting in a tight clump of individual plants over time. Vegetative spread is restricted by the shortness of the rhizomes. Plants growing in pasture have a half-life of three years. Flowers appear at a time when most meadows are mown and seed production is often prevented. Few buried seeds have been recorded in soils under pasture, although seeds in dry storage have remained viable for four years. There is no obvious seed dispersal mechanism. Regular cutting for hay reduces plant vigour. In arable land, frequent and thorough cultivation is important for control.

The Docks (*Rumex* spp., *Rumex obtrusifolius* and *R. crispus*)
The two main dock species are the broad-leaved dock (*Rumex obtusifolius*) and the curled dock (*R. crispus*). They are common throughout the UK, both as the true species and as hybrids. The hybrids may produce fewer seeds but can be more vigorous than the parents and will sometimes infest whole fields. The presence of fertile hybrids has been reported, probably the result of backcrosses with a parent. Broad-leaved dock itself is a highly variable perennial and three subspecies have been distinguished in the UK. Curled dock is capable of behaving as an annual, biennial or perennial, but plants only persist for several years when regularly cut down and prevented from setting seed. The Weeds Act 1959, requires an occupier of land to prevent the spread of broad-leaved dock and curled dock.

There are some who would argue that docks in grassland are not weeds because they contribute to the herbage and hence do not need to be controlled. They may also contribute trace elements to a grazing animal's diet. The leaves of curled dock, for example, contain unusually high amounts of zinc. However, it also has a relatively high content of oxalic acid that may affect dietary calcium bioavailability. Broad-leaved dock is relatively high in phosphate and potassium levels in the leaves, and is particularly high in magnesium. Cattle fed on the herbage containing docks are said not to suffer bloat because tannins in the dock leaves precipitate out soluble protein in the rumen liquor.

Broad-leaved dock is the most abundant dock in grassland. The curled dock is the more common dock in arable land. Dock seedlings are poor competitors and can only establish in open or disturbed patches in standing vegetation. The presence of docks in grassland is often associated with the uneven application of slurry or manure that leaves bare patches. Poor grass management, leading to overgrazing and poaching, allows dock seedlings to emerge and grow. Both docks are also found in arable crops, field margins and waste places.

Broad-leaved dock flowers from June to October but flowering is delayed by early shoot removal. The flowers are wind-pollinated but are also visited by bees. A large plant can produce up to 60,000 ripe seeds per year. The seeds become viable from the milk stage onwards, and immature seeds will continue to develop on stems cut down just a few days after

Fig 70 A dense stand of broad-leaved docks in pasture.

flowering. Broad-leaved dock can shed seed from late summer through to winter but the seeds may require a short after-ripening period before being ready to germinate. Seedlings generally do not flower in the first year. Curled dock usually flowers earlier than the broad-leaved dock. Plants may flower and set seed in the first year. Curled docks that flower in their seedling year do so from July onwards. Some plants die after flowering, while others over-winter as rosettes. The upper part of the flower panicle may be in bud, while the lower is forming fruit. Once seeds begin to form, they will ripen, even if the plant is cut down. Seed numbers per plant range from just 100 to over 40,000.

There is considerable variation in germination characteristics between seeds from different populations, different plants, different panicles on the same plant and seeds from different positions on the same panicle. Some of this is due to seed size and seed-coat thickness, some to the time of ripening and some is due to maternal factors. Defoliation can also affect seed development and germination characteristics. The seeds can germinate any time that conditions are favourable but the main flushes of emergence are in March to April and July to October. Seeds germinate best on the soil surface or in the upper 1cm (⅜in) layer of soil. A period of heavy rainfall will stimulate a flush of emergence. Germination is inhibited under a dense leaf canopy. Dock seedlings have a low competitive ability and find it difficult to become established in closed vegetation. Docks over-winter as a rosette of small dark leaves with a stout taproot.

Dock seed numbers in soil have been estimated at 12 million per hectare (5 million per acre). The seeds contain a chemical that inhibits microbial decay and are capable of surviving in undisturbed soil for over fifty years. There is no obvious seed dispersal mechanism and the seeds are often shed around the parent plant. Seeds are likely to be shed and spread during crop harvest. Seeds that have been combine-harvested, germinate more readily, probably due to scarification of the seed coat. Dock seeds can float for up to two days and have been recovered from irrigation water. The main method of long distance dispersal is as a contaminant in crop seeds, animal feed, straw and manure.

The seeds can pass through cattle unharmed and will remain viable for several weeks in manure. The seeds can also survive long periods of immersion in slurry that is not aerated. Seed viability is reduced in silage, particularly when additives are used to aid fermentation. The normal treatment temperatures that occur in sewage sludge may not kill dock seeds . Dock seeds are destroyed when fed to chickens but dropped seeds may contaminate the poultry manure. Dock seedlings have been raised from the droppings of other birds and viable seeds have been found in wormcast soil. In arable land and elsewhere, it is important to prevent the introduction of dock seeds in straw, crop seeds, manure, slurry and on machinery. In combinable crops, the aim should be to collect up dock seeds shed during the harvesting operation and denature it before disposal. Flame weeding the foliage and drilling into the roots with a drill bit or a hot spear have not given effective control of docks. Biodynamic preparations containing the ash of dock seeds do not appear to have any effect either.

In pasture, individual plants of broad-leaved dock can be very long-lived, forming compound crowns with multiple taproots. There is considerable confusion about the ability of docks to regenerate from these underground organs. It is considered that only the upper 9cm (3½ in) of the underground parts of broad-leaved dock will regenerate. Curled dock does not regenerate vegetatively as extensively as broad-leaved dock. It is thought that only the upper 4cm (1½ in) of the underground parts of curled dock will regenerate. A dock seedling takes forty days from emergence to develop a rootstock that will regenerate after decapitation. Dock plants that have been uprooted can regrow if left on the soil surface, even following a period of dry weather.

In resown grass/clover infested with dock seedlings, cutting will reduce seedling numbers initially. Frequent cutting aids seedling development and encourages regeneration of taproots and branching of the shoots of established plants, increasing the potential for future growth. Mowing has little effect on established docks but will prevent seed production. In a pasture heavily infested with docks, the best option may be to plough and reseed with grass – but not immediately. The docks are likely to regenerate both vegetatively and from seed, and a period of fallowing or arable cropping may help to reduce re-establishment. In any grassland it is prudent to avoid sward damage from trampling and poaching.

Docks are grazed off by cattle, sheep, goats and deer, but not by horses. It has been suggested that sheep should be used to graze off seedling docks in the autumn and mature docks in March to May when they are more palatable. Dock plants in and around the field should be prevented from seeding. Where seeding has occurred, shallow cultivations will encourage germination and reduce the surface soil seedbank, but only if further seed shed is prevented. Established plants should be removed by spudding, pulling or using a docking-iron when the soil is wet. This should take place once the flower stem lengthens but before flowering. The curled dock generally has a straighter taproot and is easier to uproot intact than the broad-leaved dock. All plant parts should be burned and must not be thrown on headlands or in ditches where they are likely to survive.

Undersowing cereals with clover has reduced the number of docks reaching maturity. Young seedlings can also be destroyed by thorough cultivations or ploughing. Control of established plants is by removing the docks bodily after ploughing or during bare or bastard fallowing. Potato and carrot harvesters have been used to lift and separate the dock roots from the soil for disposal off field. Ploughing followed by fallowing and repeated cultivations, during spring and early summer, exhausts the older roots and controls young seedlings. Another suggestion for the control of established docks is a series (three to four) of rotary cultivations, preferably in April to June. The rotovations begin at a shallow depth and become progressively deeper with time to around 15cm (6in). This is best done when the soil is moist, as, apparently, regeneration is less likely in wet soils, but in any case should be done with care so as not to exacerbate the infestation.

There is the potential for biological control with a range of insects and fungi. The larvae of some native weevils, *Apion* spp., bore into the flowering stems and those of a leaf-mining fly, *Pegomya nigitarsis*, can infest the leaves.

A small, leaf-feeding beetle, *Gastrophysa viridula* is another potential biocontrol agent, as it regularly defoliates dock plants. Among the rust fungi that infect *Rumex* spp. is *Uromyces rumicus*, a non-systematic fungus that can cause serious foliar injury. Natural colonization by insects and fungi may take several years to build up and become effective.

Common Nettle (*Urtica dioica* L.)

Common nettle is a rhizomatous to stoloniferous perennial, abundant and generally distributed throughout the UK. The rhizomes have difficulty penetrating compacted soil and it prefers open textured soils. Nettle is prolific on the rich land that borders meadows and pastures, often encroaching into the field. It is also troublesome around the margins of arable fields but does not spread far into the field except as isolated seedlings. Nettle is variable in size, leaf shape and flower form, and several varieties have been described.

Nettle flowers from May to September and plants bear only male or female flowers that are usually wind-pollinated. Flowering is inhibited by drought and shade. Plants cut when the seeds are at the milk stage, ripen seeds that germinate normally. Seeds are able to germinate immediately on a bare soil but germination is delayed in closed vegetation. Seeds in cultivated soil emerge sporadically through the year, with a peak in April. Plants do not flower in their first year.

Nettle has tough, yellow roots and creeping stems that root at the nodes. The horizontal shoots develop a short distance below the soil surface. New rhizomes are formed in late summer or autumn from older rhizomes or from the stem bases of aerial shoots. They continue to grow until the death of the aerial shoots and then turn upwards to form new shoots. The shoot tips may die back if frosted.

Abundant seeds are produced, most are short-lived but some seeds remain viable after five years in cultivated soil. Seeds have been recorded in large numbers in the soil beneath pastures. Seeds did not lose viability after two years in dry storage. Common nettle seeds have been found as a contaminant in samples of grass seeds. The seeds are ingested by worms and excreted in wormcasts. Seeds are also dispersed in the droppings of cattle, deer and birds.

Control is by removing the rootstocks as thoroughly as possible when nettle patches are small. Repeated hoeing will exhaust the rootstocks eventually. Seedlings may be destroyed by frequent surface cultivations in spring and autumn. The shallow, creeping rhizome does not regenerate well after repeated fragmentation. In grass, regular cutting, beginning when shoots appear in spring and repeated each time shoots reach 15–30cm (6–12in), should effectively destroy it. The regular trampling of

Fig 71 Patches of nettle in pasture.

cattle can wipe out common nettle. Salt licks around nettle clumps will attract stock to trample the weed. Livestock will readily eat cut nettles but avoid the growing plant. The 'eco-puller' has been developed to mechanically remove perennial weeds from pasture. The machine works best where there is a height difference between the grass and the weed, as the weeds are fed between rollers that pull vertically to lift out the upright stems with many of the creeping stems and roots attached. The uprooted plants are deposited in a collecting hopper for disposal.

A wide range of beneficial insects feed on the stinging nettle aphid (*Microlophium carnosum*), populations of which increase in April to May. For this reason, the time that the nettles are cut down may be important in diverting predators on to crop plants. Mid-June would appear to be the best time for cutting to allow predators to build up and then be moved on to nearby pest infestations. Cutting in May could reduce predator numbers by removing their main food source. Cutting in July may be too late to be effective. Common nettle is the main food plant for the caterpillars of several butterfly species.

Creeping Thistle (*Cirsium arvense* (L.) Scop.)
Creeping thistle is native in cultivated fields, waste places, hedgerows and grassland throughout the UK. It is an aggressive weed that occurs on most soils but it grows more extensively on deep, well-aerated soils. It is the commonest perennial weed of grassland on beef and sheep farms. It can also survive in all but the most intensively managed arable fields. Large patches of creeping thistle may be derived from a single clone but usually consist of several individuals. Defra has powers under the Weeds Act 1959 to require an occupier of land to prevent the spread of creeping thistle.

Creeping thistle flowers from July to September and sometimes into October. Normally each plant bears only functionally male or female flowers. Flowers within a patch formed from a single clone cannot self-fertilize and few seeds are set in some instances. The flowers are insect pollinated but the pollinators often visit only one type of flower. Male and female plants need to be less than 50m (160ft) apart for seed to be set reliably. Occasional plants may be found with hermaphrodite flowers and these can set seed freely. Seeds ripen from June to September and are shed from August onwards. There may be 20 to 200 seeds in each flower head and an average of 680 seeds per flower stem. The time from first flowering to seeds becoming viable is around eight to ten days. Plants cut down in flower produce very few seeds and none are viable, although they appear normal. Seeds are shed largely in autumn; some seeds can germinate on dispersal, others are dormant. Chilling over the winter leads to a peak of germination in spring but seedlings can emerge at other times too. Seeds will germinate on the soil surface but the optimum depth is 0.5–1.5cm (½ in), although emergence has been reported from up to 6cm (2½in) deep. Seeds germinate best at relatively high temperatures.

Thistle seedlings are sensitive to drought and are unlikely to survive in dense stands of other plants. A seedling at the two-leaf stage has a taproot 15cm (6in) long with spreading laterals. Adventitious buds develop at the base of the side roots allowing the seedling to regenerate if hoed off. Some of the buds will grow upwards to form leafy shoots, others develop as rhizomes. Autumn germinated seedlings may not survive if they have made insufficient root growth before the foliage is killed by frost. In spring, underground shoots that developed on the storage roots in the previous autumn, grow to the surface and develop into the new aerial shoots. Adventitious roots develop on the shoots and some swell to form the perennating organs for the following year. The food reserves of creeping thistle are minimal between May and July, in the lead up to flowering. From July to October the carbohydrate reserves build up again for the following year. The shoots die down to just below soil level in the late

Fig 72 Small creeping thistle plant in newly sown pasture.

autumn. Some of the roots may rot away leading to fragmentation of the parent plant.

Creeping thistle persists and spreads chiefly by means of the horizontal underground creeping roots that can exceed 5m (16ft) in length. The deep-seated root system is very brittle and easily broken into pieces. However, although the roots may penetrate deep in the soil, most regeneration is from roots within the plough layer. It is only the thickened areas of root and the underground stems that are able to regenerate and form new plants. Root fragments over 5cm (2in) in length regenerate readily, but pieces shorter than 2.5cm (1in) do not. Fragments of root from plants in the field margins can be carried into the field by cultivation. Undisturbed pieces of swollen root can remain dormant in the soil for several years until disturbed by cultivation.

Seeds from thistles in the headland can help to maintain an existing field population. Potentially, the seeds can be transported a considerable distance by a strong breeze but, in general, the seeds are firmly held in the seed heads while the parachute is readily detached, and seeds land within a short distance of the parent. In disturbed soil, most seeds germinate in the first year after shedding but odd seedlings will continue to emerge over the next four or five years. Seeds buried deeply in undisturbed soil can persist for over twenty years. Creeping thistle seeds have been dispersed as contaminants in various crop seeds. Seeds of creeping thistle have been found in cattle manure and can withstand submergence in water for up to two years. Viable seeds have been recovered from irrigation water.

In grassland, creeping thistle infestations may be the result of poor management. Under utilization, when thistle shoots are present, combined with overgrazing in winter and spring, produces an open sward that offers little competition to emerging thistle shoots. Close stocking or cutting the thistles at a young stage should reduce an infestation. Sheep and ponies will eat young thistle shoots readily but mature stems are not palatable to most stock. Goats, donkeys and llamas are said to eat creeping thistle, even at the flowering stage.

Where topping is carried out for thistle control, the cutters need to be set low enough to remove all of the thistle leaves. Topping must be repeated at least twice during the growing season, over several years, to have any permanent effect. In fertility building legumes, maintaining a dense crop appears to be more important than the mowing regime in suppressing creeping thistle. In roadside verges, increasing the cutting frequency reduced the occurrence of creeping thistle. Pulling is more effective than cutting and avoids new shoots simply developing from buds at the base of the cut stem. When the shoots are pulled, new shoots have to develop from the underground roots. Repeated pulling will help to drain the plant's reserves. Thistle shoots should be pulled when at the flower-bud stage.

Control through cultivation involves working the soil as deeply as practicable, with a series of operations spaced through the growing season. Deep ploughing will help to loosen the roots and bring them to the surface, where cultivators, harrows, thistle bars and the like can continue the strategy. Thistle roots and shoots brought to the surface can be left to desiccate in the wind and sun, but the roots can withstand drying down to 20 per cent moisture level. Bare or bastard fallowing is practised with good results on heavy soils. Cultivation every twenty-one days during one growing season has resulted in a 99 per cent reduction in creeping thistle. In spring barley, crop competition reduced the height, shoot density and biomass of creeping thistle plants regenerating from root fragments. Competition also suppressed the growth and spread of creeping thistle in headlands sown with grass or wildflower/grass mixes, in comparison with unsown headlands.

Creeping thistle can be attacked by rust fungi that weaken and stunt the plant, and prevent flowering. If a systemic infection could be established, a rust fungus might cause premature death of creeping thistle. Caterpillars of the painted lady butterfly can defoliate creeping thistle. It is also a food plant of some swift moths. Pre-dispersal seed predation is a major reason for low seed output by creeping thistle. The main predators are insect larvae.

Field Bindweed (*Convolvulus arvensis* L.)
Field bindweed is native in cultivated land, roadsides, railways, grass banks and in short turf throughout England, Wales and Ireland, but is rare in Scotland. It occurs on almost all soils, in many different crops, but is a particular problem in cereals and perennial crops. Field bindweed flowers from June to September. The flowers are insect pollinated. Clones are self-incompatible and differ in the timing and capacity for flowering. Seed is set from August to October and there are one to four seeds per capsule. The average seed number per plant is 600 and seeds can become viable ten to fifteen days after flowering. Seed production is greater in hot, dry summers. The hard seed coat is responsible for dormancy and damage to the seed coat results in faster germination.

Some seeds germinate in autumn and others at intervals through the year, but most seedlings emerge in late spring. A seedling may flower in its first year if conditions are favourable. Seedlings rapidly develop a vertical taproot from which lateral roots are produced, mainly in the upper 30cm (11in) of soil. The lateral roots grow out horizontally for up to 75cm (29in), before turning down to form secondary vertical roots. These give rise to more laterals that again turn down to form verticals and so on. Roots may reach 4m (13ft) deep after two and a half years. New shoots form mainly at the point where laterals turn down vertically. Apart from the seedling shoot, all other shoots originate from root-borne stem buds that give rise to vertical stems or rhizomes. The rhizomes are stout but brittle and often spirally twisted. The majority of the roots and rhizomes are in the upper 60cm (23in) of soil. The above-ground shoots appear in May and persist until the hard frosts. Field bindweed over-winters by means of roots and rhizomes. Roots in the upper layers may be killed by a penetrating frost and most lateral roots die back each year but some persist and spread horizontally.

Most seeds are hard coated and can lie dormant in soil for more than twenty-eight years. The seeds can survive for several months in silage and manure. Seeds can remain viable in the stomachs of migrating birds. Initial spread may be by seed but, once established, vegetative spread is more important. Field bindweed spreads mainly as fragments of rootstock that are able to produce new plants. Field bindweed plants at the edges of arable fields rarely spread far into the field.

The depth of the root system makes it impossible to control by cultivation alone. Only by exhaustion and removal of the rootstock can field bindweed be eradicated. In field crops, this entails short rotations with

Fig 73 Field bindweed, a pernicious weed.

extra root crops and persistent hoeing. During tillage operations, the rootstocks can be collected by harrows or by hand and these should be burnt. Turning up the rootstocks to dry in the sun during summer fallowing will reduce the weed. Sometimes only more drastic bare fallowing, with regular cultivations, will reduce it appreciably. New shoots arise within seven to fourteen days of shoots being hoed off. Field bindweed often responds to injury by producing more shoots than were originally cut back. Repeated cultivations over a period of two to three years may eradicate the weed. The optimum interval between cultivations is considered to be twelve days after regenerating shoots emerge. The longer the interval, the more prolonged the period before control is achieved. However, cultivating every two weeks initially, and every three weeks later in the year, is more practical. The only benefit of deeper cultivations is that shallow ones require a shorter interval between operations.

It is important to hoe off seedlings soon after they emerge. Most twenty day old seedlings (four leaf stage), cut off just below the soil surface, are able to regenerate after decapitation, but younger seedlings are less likely to regrow. Seedlings over six weeks old are unlikely to be killed by shallow cultivation. In arable situations, a dense crop stand may out-compete bindweed seedlings. In perennial crops, the period before planting is the main opportunity to deal with the weed. Field bindweed is not common in grassland and is unlikely to appear in closely grazed pasture. When it does occur, harrowing in spring may help to keep it down. Sheep and cattle eat the foliage, and pigs and chickens may unearth and consume the underground stems and fleshy roots.

Field bindweed seed is moderately susceptible to soil solarization. Entire or woven black plastic, or other fabric sheeting, will suppress field bindweed emergence but the cost can only be justified in long-term or high value crops. Biological control with fungal pathogens has been investigated but timing and weather conditions are critical if biological control is to be effective.

Perennial Sow-Thistle (*Sonchus arvensis* L.)

Perennial sow-thistle occurs widely on arable land and waste ground throughout the UK. It prefers slightly alkaline to neutral soils and does not thrive in acid or highly alkaline soils. Flowering takes place from late July until early October. The flowers are insect pollinated and more or less self-sterile. A few seeds become viable just four days after flowering and full seed maturity is achieved after ten days. Seeds are able to develop if the stems are cut down and left to dry, nine days after flowering. An average of forty-six seeds are produced per flower head. Seed number per plant is around 13,000, but a single flower stem with many heads may have over 9,000 seeds.

The main period of seedling emergence is March to May, with a peak in April. The seeds exhibit only a short period of dormancy and readily germinate when conditions are favourable. Most seeds germinate at 0.5–3cm (⅖–1⅕ in) deep in soil. In favourable conditions, a seedling may flower in its first year. The aerial shoots do not survive the winter, the foliage and fine roots die in September and plants over-winter as thickened roots and the underground stems of aerial shoots. In the autumn, the roots develop a strong innate dormancy that is not broken by tillage. Shoot development from anywhere along the roots begins again in the following spring. Stems begin to elongate in mid-June and flower buds are visible by the end of the month.

Seeds may remain persist in soil for at least six years. Perennial sow-thistle spreads both by seed and by the creeping root system. Seeds are generally wind-borne and dispersal distances of 6–10m (20–32ft) have been recorded in light winds. Seeds have been found as a contaminant in home saved cereal seed. Seeds ingested by earthworms have been recovered intact and viable in wormcast soil. The vegetative spread of perennial sow-thistle is due to the development of shoots that arise from buds on the swollen roots. The roots appear to live for at least two years and are found mainly in the top 15cm (6in) of soil. Radial extension of the roots of 2–3m (6–10ft) per year has been observed. Thickened roots that have reached 0.1–0.15cm (½ in) in diameter are able to regenerate when broken into segments. Root segments less than 2.5cm (1in) long, with well-developed buds, can produce new plants.

Perennial sow-thistle is susceptible to repeated tillage in the early part of the growing season. Fragmentation induces more buds to grow on the root pieces, using up food reserves. After defoliation, new aerial shoots come mainly as lateral shoots from the basal underground parts of the cut shoots. Repeated shoot removal over a period of eighty days is said to exhaust the underground organs. Shoot regeneration from buried root sections is much less around the five- to seven-leaf stage, from late-May to early-June, when food reserves are naturally low. Regular defoliation of plants at the four- to six-leaf stage is reported to have killed plants within a season. Ploughing in April followed by cultivations at three- to four-week intervals, between June and September, has given 99 per cent control of perennial sow-thistle in the USA.

Fig 74 Perennial sow-thistle.

Laying badly infested land down to grass will crowd out the weed. Dense crops of lucerne, vetch or maize will also tend to shade it out. The inclusion of successive root crops in the rotation will help to destroy the weed. Winter cereals favour perennial sow-thistle, which has the winter, spring and summer to become established and enough time to flower and set seed before crop harvest. Control should aim to prevent seeding, as well as exhaust the rootstocks. Fallowing for one year reduced seed numbers in soil by 60 per cent. Seed numbers remained low even after the land was cropped again. Seeds of sow-thistle are often predated in the flower-head by beetle larvae.

Perennial Grass Weeds

Correct identification of perennial grass weeds is important if effective control measures are to be applied and, where there is doubt, a good-quality identification guide should be used (*see* Appendix). Rhizomatous grass weeds are often referred to as couch, when they may be creeping-soft-grass, or black or creeping bent, rather than common couch.

Black Bent (*Agrostis gigantea* Roth.)

Black bent is a native rhizomatous grass, which is a serious weed of arable land. It is distributed throughout the British Isles but is commonest in the south and east. It grows equally well in marshy or dry places. Black bent is said to prefer lighter soils than common couch and to be less tolerant of tillage.

Flowers are formed from June to August and seeds develop rapidly. A third of the seeds can be viable just one week after flowering. In the north of England, black bent sheds its seed several weeks earlier than common couch. A single panicle contains about 1,000 viable seeds. Mature seeds are non-dormant and germinate readily on moist soil. The fresh seed requires light and alternating temperatures for germination, but older seeds will germinate at constant temperatures in the dark. Seeds on the soil surface germinate better than seeds lightly covered with soil and few seedlings emerge from deeper than 25mm (1in) down in the soil. The optimum depth of emergence is 0 to 0.5cm (0–⅕ in). Buried seeds germinate when brought to the soil surface during soil cultivation. The main period of seedling emergence is from March to October.

Black bent seedlings grow rapidly and initiate rhizomes at the six-leaf stage, when the plants may already have ten tillers. Seedlings are much more susceptible to competition than plants derived from rhizome pieces. Black bent rhizomes occur mainly in the top 5cm (2in) of soil but are found down to 15cm (6in) deep. The aerial shoots can also root at the nodes.

Rhizome multiplication is considered to be the main form of reproduction but seed production is also important. Most seeds germinate during the first autumn but a few may persist and remain viable in cultivated soil for at least three years. Seeds persist longer when buried and left undisturbed. Seed that has passed through the digestive system of sheep germinates better than fresh seeds. However, after three months in manure the germination potential is just 3 per cent.

The prevention of seeding and removal of the creeping rhizomes are important in the control of black bent. Forking out may be practised on a small scale and, on a larger scale, machinery has been developed with banks of rigid soil-loosening tines and shares that tear up the stubble ahead of rotating, curved tines. These flick the rhizomes out on to the soil surface, where they can be left to desiccate or can be collected up for burning.

In cereal stubble, control can include one or two passes with a rotovator, the second when regrowth has one or two leaves. The land is then ploughed, cultivated and drilled with a spring crop. The treatment does not lead to complete eradication and needs repeating in subsequent years. In pasture, the grass should be cut before black bent sets seed. Cutting intervals of two to four weeks are needed to reduce black bent populations but the composition of the sward can also have an effect. Cutting at intervals of longer than four weeks does not reduce rhizome biomass. Black bent seems to prefer acid soils and liming may check its development. Liming is said to reduce all species of *Agrostis.*

Common Bent (*Agrostis capillaris* L.)

Common bent is a rhizomatous grass, common and widely distributed, particularly on acid grassland. It is characteristic of upland pasture in short turf and can establish quickly after soil disturbance, vegetation clearance or burning. There is wide variation in growth habit between and within populations, and there are also many different cultivars. Common bent forms hybrids with creeping bent. Common bent flowers from June to August, the flowers being wind-pollinated. Seed is set from August to October and, in the field, germinates in autumn and spring.

Common bent spreads both by seed, by short rhizomes that lie just below the soil surface and by stolons. They are much thinner than couch rhizomes. Unlike most grasses, the seeds exhibit some persistence and have been recorded in large numbers in the soil beneath pastures. Seeds in undisturbed soil retained 28 per cent viability after four years. Seeds remained viable for over seven years in dry storage.

Isolated patches of common bent may be forked out and burnt Smother crops of maize, vetches or mustard will assist in choking out the weed. Under frequent cutting, common bent produces small tillers close

Fig 75 Common bent growing alongside onions.

to the ground. It exhibits some tolerance to burning. In permanent grassland, common bent is reduced when soil fertility is increased. Severe grazing encourages common bent to appear and within five years it can make up a significant proportion of the sward. The grass should be cut to prevent seeding. Common bent is grazed by rabbits and this inhibits flowering.

Common Couch (*Elytrigia repens* (L.) Desv. ex Nevski.)
Common couch grows on most soils except those with a low pH. It prefers heavier land but is able to spread more readily in lighter soils. The UK distribution of couch closely follows that of cultivated land, although it is less frequent in old grassland and permanent pasture. Common couch can form dense stands that exclude other vegetation. If left undisturbed, a mat of young rhizomes forms in the upper 10cm (4in) of soil.

Couch flowers from May to September. Although it is self-sterile, and large patches may consist of a single clone, fertilization is not usually a problem because the flowers are wind-pollinated. The seed heads mature during August and September, at around the time of cereal harvest.

Some seeds are viable ten to eighteen days after flowering, despite their green and immature appearance. There are usually twenty-five to forty seeds per stem but not all are viable. Shed seeds may germinate at any time that conditions are favourable. In the UK, germination occurs mainly during the autumn but seedlings also emerge in spring, especially when autumn germination is delayed. Seeds readily emerge from 0 to 0.5cm (0–⅕ in) deep but few from 10cm (4in).

Seedlings begin to develop rhizomes at the four- to six-leaf stage, around the time of first tillering. In most situations, vegetative reproduction is more important than seed. The aerial shoots of the parent plant die down in the autumn and new primary shoots start to develop. These grow slowly until temperatures rise in spring, when active growth begins. New leaves are produced and dormant buds at the base of each shoot grow out to form upright tillers or horizontal rhizomes. The rhizomes themselves form numerous lateral rhizomes in July. The rhizomes grow horizontally in summer before turning erect in autumn ready to form a new aerial shoot. Cultivation disrupts the seasonal growth cycle. If a rhizome is separated from the parent plant, the axillary buds develop into aerial shoots that grow vertically upward. Regeneration will follow soil disturbance at any time except mid-winter, but is greatest between October and April, and least in June. The stem bases of aerial shoots are also able to regenerate after cultivation.

Seeds can be an important source of new infestations. Couch seeds are not innately dormant and most will germinate within twelve months of shedding. Seeds shallowly incorporated into the soil germinate more readily than seeds left on the surface. Seeds buried more deeply can lie dormant in soil for two to three years and may remain viable for about five years. Contamination of cereal and other crop seed with common couch seed was, and remains, an important source of spread. The introduction of new clones increases the likelihood of seed set. The seeds retain viability after passing through horses, cows and sheep, but not pigs. Seeds may also survive in manure and have been recovered from irrigation water.

Once couch is established, repeated cultivations must be practised to reduce it. The land should be ploughed shallowly and as much weed as possible collected by grubbing and harrowing when the soil is dry. It can be almost completely killed in one season by repeated cultivations that begin in spring. The best indicator of when tillage should be repeated is when regrowth has reached the three- to four-leaf stage. In a fallow period, progressively deeper spring-tine cultivations should aim to bring rhizomes to the soil surface to desiccate them. Actively growing rhizomes are readily killed when exposed to dry air for a few days at moderate temperatures. However, if covered, even with a shallow layer of dry soil, the

Fig 76   A serious couch infestation in onions.

rhizomes may survive. The best time to work the land is when the soil is beginning to dry and falls readily from the rhizomes. In drought conditions, the rhizomes are less susceptible to desiccation because growth is restricted and the rhizome buds become dormant. Pigs, in a moveable enclosure, will root out and consume the rhizomes. Cattle and horses are also said to relish the rhizomes. Geese will eat common couch and may be selective in certain crops.

Repeated cultivations are not good for a poorly structured soil but a full fallow should not be needed on light land. A bastard or half fallow can precede fodder or vegetable crops in spring or ploughing can be delayed following a forage crop or early cereal harvest. In cropping systems without a fallow period, apart from repeated inter-row cultivations, the main period for couch control is after harvest. In cereals it is critical that rhizome fragmentation begins straight after harvest. The first cultivation with a rotovator working to 15cm (6in) should aim to cut the rhizomes into short lengths. Each fragment will develop a new root and shoot at one node and a further rotovation after two to three weeks will kill many of these. Just two cultivations may be sufficient on light soil but up to six may be needed on heavy land. Ploughing to 30cm (12in) will bury short rhizome fragments down to a depth from which some will not be able to emerge.

Cutting the aerial shoots of regenerating rhizome pieces at weekly intervals inhibits further rhizome production but less frequent cutting does not.

Competition from the crop can enhance the control of couch weakened by burial or fragmentation but, in general, smother crops alone have less effect on couch growth than cultivations. Couch is sensitive to shading, however, and when continually shaded, the grass gradually dies out. Seedlings of couch are more sensitive to crop competition than regenerating rhizome fragments. It has been said that if land is laid down to grass, common couch will be eradicated within three years. If a suitable mixture of grasses and white clover is sown and efficiently managed for a few years, the weed will be gradually suppressed. Couch will not persist under a system of close grazing.

Creeping Bent (*Agrostis stolonifera* L.)

Creeping bent is a stoloniferous grass that is native in damp arable fields and grassland. It is abundant throughout Britain. Creeping bent flowers from July to August and the flowers are wind-pollinated. Seeds are set from August to October.

In field soil, creeping bent seeds show no particular emergence period. The seeds germinate whenever conditions are favourable and most persist for less than eighteen months. Vegetative spread with long, creeping stolons leads to the formation of large clumps made up of a single clone. In favourable conditions it can spread rapidly, forming dense mats that smother other plants. Creeping bent usually remains green throughout the year.

Small infestations may be dug out to prevent further spread. Field cultivations with harrows should aim to gather up the roots and creeping stems. Detached shoots can re-root following soil disturbance in arable fields. In roadside verges, frequent cutting encourages the incidence of creeping bent. In under grazed pasture, the taller growing grasses suppress it.

Creeping Soft Grass (*Holcus mollis* L.)

Creeping soft grass is a rhizomatous grass native in woods and hedgerows. It is found in open grassland, mostly on acid soils. It occurs on moist, freely drained soils, normally light to medium texture and high in organic matter. On grassy heaths it is often in association with bracken. Creeping soft grass occurs at the edges of arable fields and is a troublesome weed on light, acid arable soils.

Creeping soft grass flowers from June to July and sets seeds from July to September. The flowers are wind-pollinated. It forms hybrids with Yorkshire fog (*H. lanatus*). Creeping soft grass has rhizomes that occur at a depth of about 5cm (2in) in soil, sometimes deeper. Rhizome growth occurs in the period from May to November but is greatest from mid-June to mid-July. The rhizomes have many dormant buds that do not develop unless the rhizomes are disturbed and then fresh aerial shoots may arise from the broken fragments. New shoots are produced mainly in the autumn. The young shoots are able to over-winter.

Creeping soft grass is spread by seeds and through fragmentation of the rhizome. In natural habitats, the spread is mainly vegetative. Rhizomes generally survive for seven to nine years and can extend by 15cm (6in) per year. A clonal patch may reach 100m (300ft) across! The plant also spreads by tillers that develop along prostrate shoots.

Once established, creeping soft grass can be difficult to eradicate but does not persist under heavy grazing, being slow to recover. It survives moderate treading and disturbance. Growth becomes more luxuriant when soil fertility is increased. In stubble, thorough cultivations to fragment the rhizomes should be carried out soon after crop harvest. Rotary cultivations are best for this. Subsequent regrowth should be killed off by further cultivations at two- to three-week intervals. Ploughing to 30cm (12in) will bury the rhizomes and cover them with a good depth of soil.

Onion Couch (Arrhenatherum elatius ssp. *bulbosum*)

Onion couch occurs throughout the UK, both as an arable weed and as a component of semi-natural grassland. It exists in several forms, differing mainly in the amount of swelling of one or more basal internodes. The non-weedy form, without the swellings, is commonly known as tall oat-grass. In semi-natural grassland, both bulbous and non-bulbous forms occur together but here the bulbous forms are not as extreme as the arable weed. Plants in these populations do not breed true; offspring are likely to be a mixture of weedy and non-weedy forms.

Onion couch flowers from June to July and beyond. The flowers are wind-pollinated and self-incompatible. An abundance of seed is produced, which are shed soon after maturing and are generally non-dormant. They germinate mostly in August and September but continuing into December. The optimum depth for seedling emergence is 0.5–4cm (⅕–1½ in), the maximum being around 13cm (5in).

Aerial shoots are formed from March onwards. Stem elongation begins in April and peaks in July. Moderate frosts do not harm the shoots. In the bulbous form, the 'bulbs' begin to develop in November from the lowest internodes, which swell and grow into bulbs, 0.7–1.2cm (3⁄10–½ in) in diameter. The bulbs are formed in order, one on top of the other, and by April each shoot has one or more bulbs, and by June shoots have four or five.

Long days and higher temperatures hasten bulbing. Young bulbs are small and white but become larger and turn brown as they mature. They usually form under the soil surface but, if the crown is shallow, the bulbs may project above ground and turn green. Each bulb has a regenerative bud that sprouts to form adventitious roots and a short rhizome that becomes an aerial shoot as it nears the soil surface. At this point a crown develops, further shoot buds and roots are initiated from the crown, and new bulbs are formed above the crown.

The seeds are spread by the wind. Onion couch does not form a persistent seedbank and in cultivated soil most seeds germinate in the first year after shedding. Few seedlings emerge in the following two years.

The weed tends to become established as seedlings in winter wheat from seeds shed in the hedgerow or in the previous crop. The 'bulbs' make effective propagules in arable land. They are easily detached and spread throughout the soil profile by cultivations. Each possesses a bud that can develop into a new plant. The bulbs are capable of withstanding drought and, if buried, can remain viable for two years; but bulbs that have sprouted once, cannot do so again. Onion couch in grassland tends to be very persistent but it is not necessarily competitive or invasive and often remains very local.

There is some evidence that the bulbous form is at a disadvantage in grassland, especially when regularly defoliated. It regrows readily after the first defoliation but after subsequent defoliations, it rapidly declines. In pasture it is favoured by under grazing. In roadside verges, increasing the cutting frequency reduced the frequency of onion couch. The bulbs are very resistant to drought and can survive long periods of exposure and desiccation. The resistance to drying out means that onion couch is less affected by rotary cultivations than common couch. Traditional methods of control have been based on deep autumn ploughing, fallowing or by sowing grass and maintaining a programme of grazing and mowing.

An increase in onion couch on heavy land was associated with a switch from ploughing to minimum tillage. In spring barley, ploughing gave the greatest reduction in onion couch. Cultivation prevents seedling establishment. Once the weed is established, repeated ploughing, grubbing and harrowing must be practised. As much of the weed as possible should be collected up by harrowing. Care should be taken not to break up the string of bulbs, if possible. Early suggestions for control included skimming off and burning the surface soil. A short rotation with extra root or hoed crops will help to combat the weed but a bare fallow may be needed on heavy land. Smother crops may be used to out-compete the weed.

Other Perennial Weeds

Bracken (*Pteridium aquilinum* (L.) Kuhn.)
Unusually for a fern, bracken is a very successful colonizer and is widespread throughout the UK. Originally a lowland and woodland plant, it now infests many upland areas. The vigorous growth and dense foliage shade out other vegetation, aided by the build-up of a deep litter layer. Young rhizomes cannot tolerate waterlogging but improved drainage has allowed bracken to colonize formerly wet sites. The upper limit for bracken growth on hills is due to the effect of frost and wind on the fronds.

Bracken has an extensively branched rhizome system buried 10–45cm (4–17in) deep. Individual rhizomes have a limited lifespan of just a few years, but the system is very persistent. The rhizome system consists of thick storage organs that run deep underground, with thin shallower rhizomes on which the leafy fronds are borne. Fronds generally emerge in

Fig 77 Bracken growing in field margins.

May and spores ripen from July to August. A frond may produce several million spores but sexual reproduction requires precise conditions. The fronds die off with the autumn frosts.

Bracken can cause poisoning in stock animals, although sheep and cattle normally avoid it. Unusually, bracken is also considered a human health hazard due to the carcinogenic spores. Bracken has been used for animal bedding, as a covering for potato clamps and as a source of potash for glassmaking. The fronds make good compost for use as a soil improver. There have also been suggestions that bracken could be harvested as a renewable energy source.

The use of a 'bracken bruiser' to crush young bracken fronds is one control technique. Bruising causes the stems to bleed and this weakens them. It is best carried out between late May and early August. Cutting is less effective because the cut surfaces heal rapidly. Bracken reaches the most vulnerable stage for cutting around the beginning of June. A second cut should be made in mid-July. Cutting should be repeated twice each year for at least three years. Pulling is effective but is little used. Even when control measures are sustained for several years, complete eradication is unlikely. Ploughing in May, June or early July gives good control of bracken, especially if arable crops are grown for two successive years. If sown down to grass immediately, there may be regrowth of the bracken, especially after ploughing in winter when the rhizomes are dormant.

The rhizomes may be exposed to frost by removing the protective layer of leaf litter. Burning the dead fronds *in situ* releases nutrients back into the soil and stimulates subsequent bracken growth. The rhizomes themselves are resistant to burning. Restricted areas of bracken have been cleared through trampling by sheep or cattle during the dormant season or at early growth stages, which destroys the young shoots as they unfurl. Rooting out rhizomes with pigs works well on small patches but disturbs the soil. Pigs may also be turned out on land that has been ploughed to expose the rhizomes but an alternative food source needs to be made available to avoid the animals eating only bracken. Biological control has been tried in the UK but was unsuccessful.

Field Horsetail (*Equisetum arvense* L.)

Field horsetail is widely distributed throughout the UK in meadows, gardens and on wasteland. It grows strongly on arable and grassland but is a particular problem in perennial crops and in nursery stock. It flourishes on damp soil. Silica deposited in the stems gives them a rough, abrasive texture. The plant is toxic to sheep, cattle and horses, being poisonous in both the green state and dried in hay. The related marsh horsetail (*E. palustre*), a weed of wet, low-lying grassland, is also poisonous to livestock.

Field horsetail produces fertile, non-photosynthetic, spore bearing stems in March to April, followed by green vegetative stems in late-spring. The single cone on each fertile stem can release 100,000 spores that germinate quickly on moist surfaces. Sexual reproduction takes place only within a narrow range of conditions and, after fertilization, cell division results in the formation of a shoot apex and roots. These sporelings soon become rhizomatous and quickly develop successive layers of horizontal rhizomes at 30cm (12in) intervals, as growth continues downwards. The early stages of development are very susceptible to desiccation and few new plants are produced from spores but, once established, the plants become resistant to dry conditions.

Maximum vegetative growth of field horsetail occurs in July. The rhizome system can be extensive, both horizontally and vertically, and may reach over 1.5m (5ft) deep, depending on substrate and water table. More than half of the rhizomes are found in the upper 25cm (10in) of soil. Tubers can be produced at the nodes of the rhizomes and may be present singly or in strings of two to four. Tubers are initiated in July and formation is thought to be influence by soil conditions. Most tubers are found below 50cm (20in) depth.

Fig 78 Dense infestation of horsetail in potatoes.

Field horsetail is difficult to control by cultivation because new stems regenerate from rhizome fragments and from tubers that remain in the soil. Regeneration of single node fragments is mainly in March to May and October to November. Tubers germinate when separated from the rhizome system and can remain viable for long periods in soil. Tubers that remain attached to the parent rhizome do not germinate.

Black plastic sheeting has been found to kill rhizomes in the upper layers of soil; however, the emerging vegetative stems can penetrate some woven polypropylene mulches. Horsetail can survive periods of flooding and burning but may be sensitive to dry conditions in competition with other plants. Control measures on arable land include soil drainage, liming, deep cultivation, the improvement of soil texture and repeated cutting of vegetative and spore bearing shoots. In grass, regular mowing over a period of years may eliminate horsetail. Horsetail is unable to compete well with tall crops as the lack of functional leaves make it intolerant of shading.

Rushes (*Juncus* spp.)

Rushes occur mainly, but not solely, on poorly drained soils of low pH. Infestations often arise in disturbed areas of pasture or where the sward is weak. Rushes are a particular problem when poor pasture is ploughed for cropping. Some species have a spreading habit, others are tuft forming. Only a few of the British species are of importance as weeds and in many situations rushes are valuable constituents of the natural vegetation. Rushes provide cover for wildlife, especially wading birds.

Soft rush (*J. effusus* L.), one of the main weedy species, is widespread and forms tussocks that extend by means of the short, creeping rhizomes from which new shoots and ultimately new plants arise. Undisturbed plants grow into clumps over 1m (3ft) tall but mowing or heavy trampling alters this to a uniform spread of shoots. Soft rush is abundant throughout the British Isles and is ubiquitous in moist situations and regions of high humidity. It prefers an open situation but can grow in partial shade.

Soft rush flowers from June to July in the south and July to August in the north. The flowers are wind- or more rarely insect-pollinated. Capsules contain an average of eighty-two seeds and a plant may produce 700,000 or more seeds. The seeds are not ready to germinate until the April after shedding. Light and moisture are required for germination. Initially, seedlings are susceptible to drying out, shading and mechanical damage, but once established, they become more resistant. The rhizomes that develop form a dense horizontal mat 0.6–5cm (¼–2in) below the soil surface. Shoots commence vigorous growth in March.

In soft rush, vegetative spread is the primary method of reproduction and spread. It may form extensive clonal patches. The wind-dispersed seeds are blown a short distance away from the parent plant. The seeds become mucilaginous and stick together when wet. Soft rush seeds may remain viable in soil for sixty years.

Hard rush (*J. inflexus* L.), the other main weedy species, is native in marshes, dune slacks, wet meadows or by water on neutral or base rich soils. It is common through most of the British Isles. The hard rush is a tuft forming species with the shoots borne on an extensive rhizome system. If grazed, the shoots are said to cause poisoning in sheep and cattle. It flowers from June onwards. The average seed number per capsule is sixty-seven and there may be over 200,000 seeds per plant. Seeds of hard rush are dispersed by wind and rain splash, also on the feet of birds and on shoes.

The compact rush (*J. conglomeratus* L.), is a rhizomatous, tuft forming rush with the shoots borne on underground stems. This perennial rush is native in marshes, dune slacks, wet meadows or by water. It occurs throughout Britain on neutral, base rich and acid soils, although the impression is sometimes given that compact rush is restricted to the latter. In wet meadows and pastures it occurs on less heavy soils poor in nutrients. Compact rush flowers from early May to July and a plant may produce 500,000 seeds.

The toad rush (*J. bufonius* L.) is the most important annual species of rush and is common throughout the UK. It is native in all kinds of damp habitats both natural and artificial. Toad rush is found on tracks liable to temporary flooding. It flowers from May to September and fruits from July to October. The flowers are cleistogamous and there are around 100 seeds per dehiscent capsule. Seed numbers per plant range from 5,300 to 34,000. It germinates from April to December, with peaks in spring and autumn. Light is required for germination. The seeds can remain viable in soil for twenty years or more and have been recorded in enormous numbers in the soil beneath pastures. In a study of the annual percent decline of toad rush seeds in cultivated soil there was no measurable loss of viability with time. Toad rush seeds have been found in cattle and horse droppings.

Rush seeds require light for germination and seedlings are only likely to emerge in open areas left bare due to poaching. Rushes do best on wet soils so improved drainage will help with any control measure. Seedlings are sensitive to competition and to moisture deficit but become tougher once established.

Control of rushes is limited to cutting, grazing, cultivation and drainage. Cutting before flowering may help to stop the spread by preventing

Fig 79 Rushes in flower.

further seed shed but there may be many seeds present already in the soil seedbank. Topping in early August followed by grazing with hardy breeds of cattle or ponies over two years is said to give good control in uplands. In lowlands, cutting followed by grazing is effective. In lowland areas that flood naturally, topping followed by flooding is effective on wet grassland. Pulling of clumped rushes is another method of control. In an established sward, regular annual mowing for hay reduces compact rush. Grazing may also help but is unlikely to be effective alone. Goats will eat soft rush in grassland. Neither cattle nor sheep grazing alone has much effect on hard and soft rush.

All rushes are controlled by ploughing or rotary cultivations but this may stimulate seed germination and the establishment of new infestations, unless the new sward develops rapidly. Where an uncut field of rushes is ploughed, the rushes may reappear between the furrow slices. The vegetative tillers grow out into the light and become re-established.

A preliminary rotary cultivation may be needed to break up the tussocks before ploughing. Plants left on the soil surface, exposed to sun and wind, are susceptible to drying out; fourteen days exposure is usually fatal. Sowing to a short-term crop, followed by further surface cultivations, may be best before establishing a long-term sward. However, it may not be possible to plough on shallow upland soils.

VOLUNTEER WEEDS

Crops that persist as weeds are a particular problem because they can act as a green bridge that allows pests and diseases to persist and can therefore nullify the effects of crop rotation. The most frequent volunteers result from crops that are grown for their seeds but can also result from green manure crops. For instance, clover seed can survive for long periods in soil and can emerge and become competitive in crops. If seeding has occurred in the ley, grass volunteers can also be numerous. Many of the novel crops grown for seed in the UK, such as borage (*Borago officinalis*), evening primrose (*Oenothera biennis*), lupin (*Lupinus alba*) and gold of pleasure (*Camelina sativa*), have the potential to become volunteer problems.

Managing Volunteer Weeds

With any crop grown for its seed, the aim must be to minimize seed loss at harvest. Where seed shedding is unavoidable, a delay in post-harvest cultivations will usually enhance seed losses through germination and predation. Seedlings that are encouraged to emerge in autumn may be killed by frost or cultivation. A stale seedbed may be advantageous in reducing the volunteer problem.

Crops grown for their vegetative organs can become highly competitive volunteer weeds because of the available food reserves. Volunteer narcissus and tulips growing from bulbs are difficult to control by cultivation and may require hand digging. Where a ley has been ploughed up, clumps of pasture grasses may remain to regrow in the following crop.

Soft fruit and other perennial crops can add substantial numbers of seeds to the soil and can be a problem in the current crop and a potential problem in subsequent crops. Strawberry seedlings have emerged five years after the last crop was grown. Seedlings of blackcurrants and raspberries are common in established plantations. Raspberry seeds may make up over 25 per cent of the soil seedbank in established raspberry plantations. The seeds can also be transported to other areas by birds.

Seedling asparagus can be a problem in asparagus beds but only the female plants set seed. The problem can be avoided by selecting the male crowns or planting one of the newer cultivars that do not produce seed.

Volunteer weeds are not always a problem. Seedlings of field beans can emerge in large numbers in following cereal crops but this can be an advantage where the cereal is cut for silage. Where the crop and volunteer have similar dates of maturity, it may be possible to harvest them together and separate out the seeds.

Common Volunteer Weeds

Volunteer Cereals (Various Species)

Volunteer cereals arise mainly from grain shed at, or before, crop harvest. An average of 16 per cent of grain can be lost or shed at this time, but some cultivars are more resistant to shedding and lose just 5 per cent. Seedlings can also develop from cereal grains or whole ears in the straw used as mulch or as animal bedding. Seeds can also be dispersed by farm machinery. Grain for drilling has been found to contain a small percentage of other cereals. Volunteer cereals can be a problem when they emerge in a different cereal or in a different cultivar of the same cereal. The volunteers can result in too high a plant population leading to yield depression and other problems.

Prolonged persistence of cereal volunteers is unlikely, even if seed is deeply buried. However, some cultivars have higher dormancy levels and their seed may persist longer. Temperatures during seed ripening can also influence dormancy level, as seeds that ripen at higher temperatures and in dry conditions are less dormant. A cool, wet period at, or around, harvest may encourage greater dormancy.

Careful choice of varieties with some resistance to shedding can help to reduce the volunteer problem. Ploughing down stubble to deeply bury cereal seed is considered by some to be the best management option. Although cereal emergence is reduced, the seed has potential to persist and germinate after ploughing the following year. The emergence of volunteers is greater in reduced tillage systems, where the seeds are retained near the soil surface. Dormancy in freshly shed cereal seeds can restrict early germination and may limit the effectiveness of post-harvest control measures. Stubble cultivations can be more effective if delayed to allow the seed to germinate, so that the emerged seedlings can then be killed by the cultivations. However, the timing of stubble cultivations to encourage the germination of shed cereal seed may conflict with the need to control some other weeds.

Volunteer Oilseed Rape (*Brassica napus* L. ssp. *oleifera* (DC.) Metzg.)

Volunteer oilseed rape seedlings result from the significant seed losses that occur in spring and winter oilseed rape crops, both before and at harvest. Shed seed numbers often reach 10 million per hectare (4 million per acre). Volunteer oilseed rape seedlings can persist and have emerged nine years after the original oilseed rape crop was harvested. In these circumstances, volunteers can be a particular problem in subsequent oilseed rape crops.

Timely harvesting will reduce, but not eliminate, seed shed. The optimum stage for oilseed rape harvest is when seeds in the bottom pods are dark brown, those in the middle pods are reddish brown and seeds in the top pods are green but starting to turn brown. The seeds of oilseed rape are not dormant when shed and only develop dormancy after exposure to certain environmental conditions. The germination of freshly shed seed can be maximized by keeping seeds in the light on the soil surface and exposing them to alternating temperatures. Cultivations that bury seeds deeply in soil increase the risk of inducing seed dormancy and hence persistence. Incorporation by tillage should be avoided or delayed for as long as possible. A two-week delay in cultivation should be sufficient to encourage seed germination, providing conditions are wet after harvest.

Fig 80 An impressive infestation of volunteer oilseed rape.

Volunteer Potato (*Solanum tuberosum* L.)

Volunteer potatoes arise from seeds, tubers and tuber pieces that remain in the soil following an earlier potato crop. After harvest, as many missed tubers may be left in the soil as were planted originally. Volunteers that develop from these ground-keeper tubers grow vigorously and are difficult to eliminate in poorly competitive crops. Volunteer potatoes emerge from late-April to late-July, depending on the prevailing soil conditions and the vigour of the crop.

Where a berry-producing cultivar has been grown, volunteer potatoes also develop from seeds. Each potato berry may contain 200 to 300 seeds. The seeds can remain viable in soil for three to nine years. Potato seedlings emerge from May to late-June and can continue to appear until September. The seedlings that emerge up until June may produce small tubers that remain in the soil as ground-keeper potatoes.

More tubers will survive following ploughing, compared with shallower cultivations that leave tubers near the soil surface, where they can be killed by exposure to frost. Potato tubers need prolonged exposure at –2°C (31.8°F) to be reliably destroyed. Even in a moderately hard winter, frost is unlikely to eradicate the entire volunteer population. A strong competitive winter wheat crop can significantly suppress the growth of volunteer potatoes. The winter cereal develops a dense canopy before the

Fig 81 Volunteer potatoes in carrots.

potatoes emerge. In vegetable crops hand rogueing may be the best option, if they are not removed in normal mechanical weeding operations.

Weed Beet (*Beta vulgaris* L.)
Weed beet first became a problem where sugar beet crops had been grown in close rotation. A serious infestation can build up over three cycles of sugar beet cropping. Weed beet can originate from native wild beet, from hybrids between wild and cultivated beet, and from bolters in open-pollinated cultivars. Bolting of beet is partly inherent, the annual character being dominant, and partly under climatic control. Early sowing of sugar beet can expose crop seedlings to cooler conditions that promote bolting. Roots left in the ground after sugar beet harvest can flower and set seed in following crops. The resulting seed may produce seedlings with the potential to become annual rather than biennial plants. Weed beet can therefore evolve from seed of any beet left to flower in the field.

Bolters become visible in late May and flower in July. Individual flowers on the inflorescence open over a period of three to four weeks. The flowers produce large amounts of wind-blown pollen. Seeds start to become viable in mid-August, twenty-eight days after the start of flowering. The flowers on cut down stems or pulled plants can produce a small number of viable seeds. Intact bolters left to grow to maturity produce an average of 1,000 to 1,919 seeds per plant. Weed beet begins to shed its seed in September. The seeds appear able to remain viable in soil for at least seven years. Burial through ploughing may prolong the life of the seed.

Weed beet is less of a problem in competitive crops like cereals and in crops that are harvested early in the season. To limit weed beet infestations in sugar beet, grow the crop on clean land, use bolt resistant cultivars, sow the crop after mid-March and destroy bolters in July to August to prevent fresh seed entering the soil seedbank. Leave shed seeds on the soil surface to germinate, die or suffer predation. Shallow cultivation will stimulate seed germination. In the growing crop, it is difficult to distinguish weed beet seedlings from sugar beet seedlings, unless they are out of line with the crop row. Tractor hoeing of the inter-rows will remove 70 per cent of the weed beet. The tall flower-spikes of the bolters allow height selective control. Good control is achieved by cutting down bolters three times, at two-week intervals, starting fourteen to twenty-eight days after flowering. Two cuts will give reasonable control but a single cut does not give a reliable result. Cutting forty-two days after flowering is not satisfactory as viable seed is already formed.

Chapter 6

Weed Management Strategies in Systems and Crops

In the previous chapters we have looked at the principles of weed management in organic farming systems, discussed the practical approaches that can be taken, the economic context in which they should be undertaken and the weeds themselves. We have also stressed that weed management on organic farms is, above all, about taking a flexible and pragmatic approach over the entire rotation. As a final step, and in order to make up a farm strategy, all this information needs to be put together in a manner consistent with the farmer's intentions and goals. Normally this would necessarily also be consistent with guaranteeing them a sustainable living within the framework set by organic principles.

This chapter aims to bring together some of these elements and discuss them at two levels: that of individual crops within rotations and at the farm level over the whole cropping cycle. There is clearly not sufficient space in one book to cover all possible crops and rotations. We have, therefore, opted, first, to discuss weed management in broader rotations typical of organic farming systems in temperate agriculture, especially the UK and Europe, and then to provide more specific information on weed control in broad crop groups. We hope that this, combined with a reading of the earlier chapters, will provide a broad brush canvas, which will serve to indicate how farmers and growers can think about their weed management strategies and how to proceed. Ways of further organizing and developing this information are discussed in Chapter 7 and the Appendix. Of late, much detailed information is also becoming freely available on the internet and we would recommend this as a resource and medium of information exchange, despite the frustrations that often go with this type of investigation!

WEED MANAGEMENT WITHIN FARM SYSTEMS

Farmers' and growers' attitudes to weeds can range from ambivalence to hatred, but often seem to soften (a bit!) with length of time that they have been organic. All farmers and growers have at least some strategies in mind when it comes to weed management, even if they are only very sketchily developed. Cultural controls (as described in Chapter 2) are generally very important within organic farming systems but the details depend on the specific farm system. Most farmers and growers will also have a few favourite direct control methods, which they can deploy rapidly within a season or as a situation develops (as described in Chapter 3).

It is striking that all farmers and growers can record numerous success stories of managing weeds within their farm systems, and even where they record failures, they can explain, at least in part, why they have failed and what can be done to improve the situation. Recorded successes have included: manual dock removal; different sequences of operations for dock control; topping weeds at the right time; adapting kit for use on the farm; using break or fertility building crops, like red clover or potatoes, to suppress weeds; the use of different varieties; the use of sheep and pigs as part of weed management regimes; and learning to live with weeds. In fact, as many successes as there are farms. On the other hand, failures are often put down to inexperience, lack of time or lack of options. Particularly difficult areas include; dock control; headland management; crop emergence; pest attacks causing weed problems; and missing weeding windows.

Here we give an overview of weed management in the context of typical organic arable, grass and vegetable systems of the type that are encountered in temperate organic agriculture. All decisions should reflect on whether weed control is needed in the light of their likely effect on marketable yield, as well as their likely negative effects in the short-or long-term. This should be contrasted against any beneficial effects they might be having. Remember that removing all weeds is difficult (or even impossible!), is not always desirable and does not necessarily make economic sense.

Arable (and Cereal) Systems

A dense stand of cereal may appear relatively competitive against weeds but, at early crop growth stages or under unfavourable conditions, particularly where crop stand is reduced, weeds can take over and cause substantial yield losses. In addition to causing a loss of yield, the weeds may delay ripening, hinder harvesting and reduce grain quality. For instance, weeds such as cleavers, black bindweed and the vetches can reduce crop yields by causing lodging. Weed seeds may contaminate the grain and require additional seed cleaning or make the crop unmarketable for certain purposes.

In a comparison of cereals in organic and conventional rotations, there was a far greater range of weed species in the organic crops. The main weeds found in the conventional crops were cleavers and black grass but these were present at much lower levels in the organic cereals. Perennial weeds including dock, creeping thistle and rough meadow grass were more important in the organic crops.

Problem weeds are likely to vary with the specific conditions and cropping practices used, and different weeds can be expected to cause problems in spring and winter cereals. For instance, growing autumn cereals favours wild oats, barren brome and black grass. The peak of emergence for other weed species is the spring, and the sowing date of spring-sown cereals will affect the composition of the weed flora. With earlier drilling, late-summer germinating weeds, such as red dead-nettle (*Lamium purpureum*) and the cranesbills (*Geranium* spp.), become more frequent. Still other weeds will emerge at any time that conditions are favourable; for example, annual meadow grass, chickweed and shepherd's purse. Cropping practices will also have an effect; for example, seedling numbers of many annual broad-leaved weed species have been observed to decrease under direct drilling, while grass and perennial weeds increase. In general, earlier crop sowing appears to increase the frequency of weeds, as does the number of years that cereals are grown without a break. Soil conditions can also be important so, for instance, wild oats are more frequent on heavier soil and the meadow grasses on lighter soil.

In purely arable systems, imaginative use of fertility building periods and break crops is essential for managing weeds. In stockless systems, where there is only a short duration fertility building crop, it has been shown that crop rotation design is important in managing weeds and that growing successive wheat crops results in the most serious weed problems. Varying rotations with weed management in mind (for example, alternating winter and spring cropping; alternating cereal types) can help reduce weed pressure, as can the inclusion of potatoes in a predominantly cereal rotation, in order to break the life cycle of autumn and early spring germinating weeds. In mixed systems, that is those that have livestock, the grass ley management will be important for weed control (*see* below) as will the use of suppressive break crops, including legumes, that will also serve to supply animal feed. In these mixed systems, treatment of

Fig 82 Brome, a grass weed, in a crop of barley.

manure and slurry is also important to prevent the return of weed seeds to pastures.

Cereal species differ in competitive ability. Winter oats, spring oats and triticale are more competitive than wheat and tend to have lower weed populations. In field studies of the effect of barley, oats and wheat on cleavers grown at a range of densities, oats were the most suppressive of this weed, followed by barley and then wheat. Dense canopy structure is important for weed suppression but there is also the possibility that oat seedlings secrete allelopathic exudates that suppress weeds at early growth stages. Mixed stands of cereals (dredge corn) are grown in the belief that weed suppression is greater and the yield is higher than when the crops are grown alone. Mixtures of cereals with grain legumes, such as barley or oats with peas, and wheat and field beans, are also likely to give greater weed suppression. Weed biomass has been reduced in winter wheat inter-cropped with field beans, compared with the sole crops.

Within species, crop variety can significantly affect the outcome of weed management strategies. Modern short strawed cultivars, suitable for the combine harvester, are considered less able to smother weeds than the older taller ones, although straw height is not the only factor. In cereal variety trials, earliness of ground covering and canopy density are considered important characters that indicate the weed suppressing ability of cultivars and those with an early, prostrate, ground covering habit can significantly reduce weed biomass. Early ground covering is needed during crop establishment, when weed emergence is at its peak. Cultivars with horizontal, rather than vertical, leaves cast more of a shadow. Canopy density becomes more important in wet seasons when weed growth continues into the summer, unless shaded out by the crop. However, characters associated with improved weed suppression may not confer the ability to tolerate post-emergence mechanical weed control, as 'leafy' varieties are more prone to damage in blind harrowing.

It is important that only clean cereal seed is drilled. In the past, weed-seed contamination was responsible for increasing existing weed problems and introducing new ones. In 1970, a survey of cereal seed in drills on farms showed that 89 per cent of farmers own seed was contaminated with weed seeds compared with 36 per cent of merchant seed. Weed seed numbers were also much higher in the farmers' own seed. The 1922 Seed Regulations prohibited the sale or sowing of cereal seed containing more than 5 per cent by weight of the five specified injurious weeds: docks, cranesbills, wild carrot (*Daucus carota*), Yorkshire fog (*Holcus lanatus*) and soft brome (*Bromus mollis*). The 1961 Seed Regulations introduced a requirement for the declaration of the percentage of all weed seeds present in seed that was sold. For the production of certified cereal seed, careful attention needs to be paid to previous cropping in order to minimize weed-seed contamination with grass weeds – wild oat, black grass and sterile brome being particular problems. A contractor should be used to clean seed that is going to be resown.

The primary and secondary cultivations used to prepare a seedbed are crucially important in managing weeds in a cereal crop, especially in giving the crop a head start over the weeds. A wide range of machinery exists for these cultivations and there is machinery now available that will take stubble down to a seedbed in a single pass. The trend towards early sowing of spring and winter cereals to achieve the highest yields, gives little time for clearing the land of weeds during seedbed preparation and this may increase weed populations significantly. In spring cereals, allowing time for a stale seedbed helps to reduce weed numbers in the growing crop, although there are mixed opinions over the effectiveness of stale seedbeds. The exact nature and timing of cultivations will vary, depending on the specific situation (such as the previous crop, the soil type, the soil condition, the equipment available) but it is important to get a good seedbed to allow fast crop establishment. A well established crop might not need any further weeding operations and, in fact, many studies of post-emergence mechanical weed control suggest that, although these weeding operations reduce weed density and biomass, there is little yield benefit from weeding at this time. Some farmers feel that a cloddy autumn seedbed gives the emerging crop some protection during the winter but, in general, a level, crumbly seedbed will give the crop the best start, although the weeds will also emerge better and in greater numbers from a fine seedbed. Balanced against this is the fact that control measures are likely to be more effective with a fine, level seedbed.

Increasing the seed rate of cereals (for example, winter barley to 200–250kg/ha) has been shown to reduce the biomass of weeds but often the number of weeds is not affected. Many studies have indicated that there may be little yield benefit and the extra cost needs to be balanced against any weed control benefit. In addition to sowing rate, the relative time of crop and weed emergence can have a considerable effect on subsequent crop – weed interactions. Deeper sowing of the crop can result in delayed emergence and may cause a yield reduction due to weed competition. Soil moisture levels in the upper layers of soil at sowing are also important in determining the speed of crop emergence. One study showed that, at a given density of wild oat, crop yield loss increased, the earlier that the wild-oat emerged relative to the crop.

Within the growing crop, the effectiveness of direct weed control operations depends in part on the density and size of the weeds. The fewer the weeds and usually (but not always) the smaller the weeds, the better the level of control. Unfortunately, it is difficult to predict the effects of any likely weed population far enough in advance to be certain there will be an economic benefit in carrying out additional direct control measures. To determine the value of applying weed control strategies, systems of thresholds have been developed that were originally intended for use with herbicides but can be applied to any weed control input. In practice, the outputs of these models have proved of limited value in practical situations and farmers will have to rely on judgement and experience (*see* Chapter 4). As a rule of thumb, it is important to keep weeds at a manageable level using a mixture of indirect control strategies and 'good housekeeping'. It is also important to try not to recreational weed, that is, only weed if there is a justifiable reason, either an increased economic yield or a reduction in weed pressure in another part of the rotation. Walking crops and evaluating weed flora and density, before weeding, can help to provide experience of effects and can pay off in the longer term.

Blind harrowing with chain or spring-tined harrows can be used to control weed seedlings that have emerged or are about to emerge before the crop. In one study, a tine weeder and chain harrow have been show to be equally effective in giving a 40–70 per cent reduction in cereal weeds. On soils that develop a surface crust, stronger teeth are more successful. Low selectivity between crop and weeds during early post-emergence harrowing, limits the possibility of achieving a high level of weed control without serious crop damage. Field studies with a flexible chain harrow passed up to three times through a spring barley crop at the second-leaf stage, to simulate different intensities of harrowing, resulted in a range of weed control levels up to 90 per cent. However, only 21–51 per cent of the potential yield gain was achieved, due to greater crop injury as weed control levels increased. Practically, many farmers try to weed winter cereals before mid-December (although weather and soil type determine success) and carry out blind harrowing five to six days after drilling or two to three weeks after germination. Spring-sown cereals are also weeded as early as possible, with some farmers weeding all one way and others across the field and on the diagonal, in order to catch weeds from different angles.

Spring tine-weeders are probably the most widely used form of mechanical weed control in organic cereals. Under the right conditions, harrowing with spring-loaded tines can continue after the cereal crop has emerged and is past the three leaf stage, in order to control scrambling weeds, but it is only effective against the young seedlings of tap rooted weeds. The best conditions for harrowing are a drying soil and a weather outlook with no rain for two or more days after harrowing to prevent weeds re-rooting. Weeding can be done up to GS 37 and up to five times but typically farmers make two passes (but some not at all).

Established weeds are, in general, controlled more effectively by inter-row hoeing but this requires the crop rows to be drilled at a suitable spacing. Inter-row hoeing may reduce weed density by over 90 per cent but crop injury can be a problem without the use of a reliable guidance system. Mechanical inter-row weeding is possible up to ear emergence, using computerized visual-imaging systems that can automatically adjust tines in-line with crop planting to eliminate driver error. The crop may also benefit from increased nitrogen mineralization caused by soil disturbance, although this effect is marginal in most circumstances. Weeding operations should be conducted at an early stage to eliminate crop competition and to avoid disturbing ground nesting birds. To facilitate inter-row cultivation, some increase in row spacing may be required but in cereals, drilling crops in 25cm (10in) rows can reduce yield. Increasing crop row spacing from 12.5 to 15, 17.5 or 20cm (5 to 6, 7 or 8in) has also resulted in higher weed fresh weight and density. Increasing the seed rate to compensate can result in greater competition between individual crop plants. Band sowing the crop seed, or using a twin-row system, can reduce the problem. Crops are often harrowed in autumn and inter-row cultivated in spring.

Hand rogueing is of value to deal with small numbers of plants of a particular weed in order to prevent a larger population developing (for example, docks, creeping thistles or wild oats). It is also important to pay particular attention to isolated patches of weeds when they occur within a field, to prevent them spreading. Undersowing is also popular in cereal crops, as it avoids a cultivation, helps suppress weeds prior to and after harvesting, and also helps to establish a ley, although it may also have a yield penalty.

Grassland and Livestock Systems

Grassland and grass–clover leys are an important part of most organic-farming systems in the UK, where up to 70 per cent of the farmed area is comprised of mixed grasses and legume leys. Grass may be managed as a short-, medium- or long-term crop, and this may determine the composition of the desirable sward species and the nature of the associated weeds. In a survey of 502 farms in England and Wales, where grass was the major crop (and at least half was permanent grass), it was found that 50 per cent of farmers thought thistles, chiefly creeping thistle, were a problem, while 40 per cent considered docks to be a problem. Thistles were mentioned more by beef farmers, while docks were highlighted by dairy farmers. Docks appeared to be associated with low potassium and high phosphate, while the opposite was true for thistles.

Weed management in grass comprises a mixture of cultural methods, cutting regimes and managed grazing. Problems can be broadly split into two: those in leys (less than five years) and those in long-term (permanent, more than five years) grass. In these situations, the weed flora is determined by the types of repeat management practices and by physical characteristics, such as altitude, aspect, steepness, wetness, liability to flooding and rockiness. Indeed, the land may be in grass because some of these factors mitigate other types of land use. The type of weed control also depends on whether annual or perennial weeds are present. In intensively used grassland, grass weeds are often more of a problem than broad-leaved weeds, and selective control is difficult.

The seed mixture for leys may include a relatively simple mixture of grasses and legumes, or may be more complex and contain a range of 'herbs', such as yarrow, that in other situations would be considered a weed. Many consider that such species add to the nutrient value of the grass, and benefit both the stock and the land. Different seed mixtures will have different weed control effects, and timing of off-take for silage and

Fig 83 A weedy grass clover ley.

hay can also help to manage weeds by, for instance, removing weeds before they set seed. A key strategy is to sow a competitive ley mix and get good initial establishment.

Good pasture management involves maintaining the condition of the sward by cultural means. Once established, weed intrusions can be managed by chain harrowing in spring and topping regularly during the growing season. Weeds may be manipulated by the timing and intensity of mowing or grazing, and successful strategies often comprise integrated topping and grazing regimes. Grazing heights are important to prevent poaching and weeds establishing on bare earth, as is frequency of topping. It may be necessary to experiment with stock, stocking levels and crops (for example, silage/haylage/hay), according to weed problems.

Grassland stocked densely and cut for silage suffers few broad-leaved weeds except docks. A diversity of grazing animals will help to manage a broad range of weeds and prevent any species getting out of hand. For instance, rotating cattle and sheep will provide a broad grazing regime. Sheep can also be tightly grazed on new leys to control troublesome broad-leaved weeds.

Other possible cultural treatments include drainage, crushing and ploughing. Manure should be composted and slurry aerated to give maximum kill of weed seeds, before being returned to the land. Perennial weeds, such as docks and thistles, may require hand pulling in established grassland and more intensive cultivations, when new leys are established. Mowing specific patches (especially creeping thistle and nettles), and management of weeds in hedge and fence lines, can be important to prevent the spread of perennial weeds. New leys can be topped before grazing to control annual broad-leaved weeds. Common chickweed can be particularly troublesome in autumn re-seeds.

Vegetable Production Systems

Organic vegetable systems are generally diverse with more or less complex rotations. They are also intensive with high value crops in small(ish) areas. Many vegetable crops are sensitive to weed competition, in at least part of the crop cycle, and can be relatively uncompetitive (*see* individual crops below). In addition, vegetable production is also driven by markets, in which consumers directly buy, handle and eat the produce, as opposed to the produce being highly processed and modified. In this situation, experience has shown that there are a large number of potential weed species, with annual weeds being particularly prominent. When weeds arise, there is low tolerance to their presence and growers like to keep their fields weed-free!

Within vegetable systems the intensity of cultivation can reduce the importance of perennial weeds like dock, in most situations. However, the value of the crop, and the uncompetitive nature of many vegetable seedlings, increases the importance of annual weeds. This is exacerbated by the frequent and repeated intervals of bare earth following cultivations that stimulates these weeds to germinate from the weed seedbank. Actual weed floras encountered on organic holdings are very diverse, depending on climatic and edaphic conditions, as well as the farm history, but it is safe to say that all farms will encounter a range of annual, and to lesser extent, perennial weeds that need to be managed. An important consideration with annual weeds is to prevent them completing their cycle by setting seed and returning the seed to the soil seedbank.

Identifying key times and action points within the rotation are indispensable for good weed management, bearing in mind that markets often drive rotations, rather than technical constraints *per se*, and thinking long term is crucial. Weed control is often a by-product of other operations or objectives also integral to the system, such as seedbed preparation or ridging up plants. It is also important to have the ability to deal flexibly with weed problems as they arise, either within a specific crop or across the rotation.

Realistically, a weed management programme for field vegetables will aim to maintain low and tolerable levels of weed infestation and this is likely to make weed management complex (and costly!) within these systems. Most growers and advisors would maintain that weeds need to be managed at all times and stress the importance of keeping 'on top of weeds' (often to prevent the return of weed seed to the soil). Weed management needs to be integrated with crop management at all stages to be successful and all the different methods will need to be built into a system's strategy. A successful strategy is likely to need a range of approaches over a period of time, combining longer term planning (for example, rotations, crop variety choice) with short-term reactive or direct measures (for example, harrowing, flame weeding). It is important to remember to balance the costs and benefits of weed management and to maintain organic principles (sustainability, biodiversity, etc.). Nutrient and pest and disease management are also very important in vegetable systems, and low levels of weeds may even promote, for instance, nutrient retention and natural pest control.

Given these constraints, growers will often employ a range of similar strategies and tactics to manage weeds. In establishing crops, a good, fine seedbed is essential, as this facilitates planting and any direct weed control methods. Many growers consider the use of stale seedbeds as vital to flush out weeds before establishing the crop. In addition, many crop types can be raised as bare-root transplants or modules that can be planted out once their development is well underway. This can be important in allowing for time to prepare stale seedbeds and also naturally gives the crop a head start over any weeds that develop; it creates a differential between the crop and weeds that can be exploited for direct weed control operations. It is also worth thinking about other cultural control measures, such as post-harvest cultural operations and establishment of cover crops that smother or otherwise eliminate weeds before they can set seed.

When using mechanical weed control, timing is vital and most growers would recommend that a holding have at least one or two bits of kit available for weeding. Most growers adapt at least one piece of machinery to their growing system and grow vegetables on beds of standard widths and row spacing to match this. This can prevent a lot of time being wasted in constantly adjusting machinery and it allows weeding operations to be carried out in a timely manner when weather conditions allow (especially important in wetter colder climates!). A wide range of mechanical weeders is now available, including an enormous array of tine-weeders as well as more sophisticated machines like brush-weeders and finger

Fig 84 A versatile steerage hoe can be indispensable for managing weeds in organic vegetables.

weeders. Most growers would probably aim to have a fixed tine-weeder of some description, useful for inter-row operations, together with a more sophisticated weeder like a finger weeder that adds flexibility for inter-row operations. Many of these tractor-mounted weeders also come as steerage hoes, which can be steered along the crop rows by an additional operator. These hoes can add to the flexibility available for weed control and increase the speed of any weeding operation. The soil type and growing system will determine which are the best machines to use but growers should always attempt to borrow equipment and try it out before buying.

Although costs are high, manual weeding can also be important in some crops or to 'rescue' a situation. The market prices for some crops can justify manual weeding, at least on an occasional basis. Intra-row weeds can also be difficult to remove and in such cases manual weeding might help. Flat bed weeders have been developed that allow up to eight people to be carried over the crop while weeding within crop rows, which might help to reduce costs, although the work can be laborious for the weeders, who will determine the effectiveness that can be achieved!

WEED MANAGEMENT WITHIN SPECIFIC CROPS OR CROP-TYPES

Weed management strategies within crops or crop types will vary depending on the crop and the situation in which it is being grown. This crucially includes the economic situation and, linked to this, the intended market. However, the biological basis for weed management will also depend on the agronomy of the crop; where it is grown, how it is grown, when it is grown and the crop habits or characteristics. It will also depend on the weed flora present, as described in the Chapter 5. In this section, we highlight some of the factors to bear in mind when designing weed management strategies for specific crops or crop types. The list is not exhaustive but we hope that it serves to inform farmers and growers of the underlying factors, and allow them to more effectively design management strategies for all crops.

Weed Management in Beets

Four types of beets (*Beta vulgaris*) are commonly grown in the UK, including sugar beet grown as a field crop, fodder beet grown as a field crop for its roots and foliage for animal consumption, beetroot grown as a vegetable crop outdoors or under protected conditions for the edible roots, and leaf beet, also grown outdoors or under protection for its salad leaves. Weed management practice will depend on which beets are being grown. Fodder and sugar beet are discussed below, whilst weed management in beetroot and leaf beets is introduced in the section on field vegetables below.

Fodder Beet

Fodder beet may be drilled or broadcast and, in this low market value crop, weed management will mainly be through pre-sowing cultivation to suppress weeds or use of stale seedbed techniques. The crop should be given as best a start as possible, to allow it to out compete weeds at the early establishment stage (*see* also Sugar Beet below). Pre-emergence flaming has been shown to reduce weed numbers by 34–44 per cent. However, in relatively low value arable crops, such as fodder beet, that are grown on a large scale, the cost of flame weeding will not usually be justified. Punch planting makes use of the stale seedbed technique but minimizes soil disturbance even further by dropping the seed into holes made by a dibber. This technique has been shown to reduce weed density by 30 per cent, compared with a normal drilled crop, but once again the outlay on sophisticated machinery may not bear the expected monetary returns.

Sugar Beet

Although there is currently no market for organic sugar beet, this crop usually follows a grass–legume ley or a cereal in the rotation and precedes a cereal or some other crop that will benefit from the residues of any manure application. Sufficient nitrogen from a manure or compost application is important to ensure rapid leaf development that will provide a dense leaf canopy and shade out the weeds. Crop losses of up to 95 per cent have been recorded where tall weeds, such as fat hen, were present and up to 50 per cent where low growing weeds, such as chickweed or scentless mayweed, were predominant. Weed beet may be a particular problem in sugar beet (*see* Volunteer Weeds in previous chapter).

The primary and secondary cultivations required for seedbed preparation will have a considerable influence on the weeds. However, the nature and timing of these cultivations will vary with the previous crop, with soil type and with soil condition at the time of any operation. In general, a level, crumbly seedbed will give the crop the best start but growers may want to keep cultivations to a minimum. Weeds tend to emerge better and in greater numbers from a fine seedbed than a coarse one, but control measures are often more effective on a fine, level seedbed. A stale seedbed may be prepared ten days in advance of the drilling date and the weeds killed by shallow cultivation before drilling. There is, however, a risk of the seedbed drying out, which can result in erratic germination when the crop is sown. Light harrows may be used after drilling either on the flat or to level the ridges due to drilling, but this may reduce plant stand. Sugar beet cultivars vary in their growth habit, some have an erect leaf rosette (cv. Carla) others have a more horizontal leaf arrangement (cv. Lucy). Weed seedling survival can be much less with the latter, demonstrating the importance of early ground cover establishment.

It is important to achieve a good crop stand, as it is the dense leaf canopy that shades out emerged weeds and inhibits later flushes of seedling weeds. As the crop canopy does not close fully until mid-summer, many tall growing weeds, such as fat hen and certain mayweeds, may grow above the canopy before it closes. In the UK, the optimum weeding period is between four and six weeks after 50 per cent crop emergence. In practice, weeding operations should commence at the four- to six-leaf stage and may cease at around the ten- to twelve-leaf stage. Once the optimum weeding time has been reached, yield may be depressed by 1.5 per cent for each day the crop is left unweeded, although sugar beet has some ability to recover from an early check.

Mechanical inter-row cultivation is important in early control of weeds. However, cultivation stimulates further weed seedlings to emerge. Spring tine weeders can be effective in sugar beet at low weed densities, when the soil is drying and weeds are unlikely to re-root. The crop must have at least six leaves to withstand the tine weeder but must not be so large that the leaves catch on the tines and pull the crop out. Intra-row weeds are more difficult to deal with. The torsion weeder at low intensity has proved to be relatively gentle on the sugar beet crop from the four-leaf stage. In studies with this implement, altering the distance between the tines and increasing the driving speed to give higher intensity weeding, caused more damage to the crop but gave good weed control. The working depth of the tines was 1–2cm (½ in). Dry weather is required before and after cultivation, and weeds must be smaller than the crop.

Weed Management in Cereals

Weed management in cereals has already been discussed in general terms (*see* above) but there are many different types of organic cereal crops grown, which differ slightly in their details as far as both crops and weed management are concerned.

Weed Management in Barley

Barley is often the second cereal in a rotation. The relatively early harvest date of barley can also leave time for post-harvest cultivations for weed control (e.g. couch) where necessary. In pot experiments, barley has been shown to suppress the growth of charlock that emerged after the crop and, although the charlock subsequently grew more rapidly, it was unable to overcome the initial growth difference. The best competitors against weeds are the tallest cultivars, those with the greatest stem weight or those with high, early, relative growth rates. It was also shown that covering the seeds with compost (3cm/1in depth) did not reduce barley seedling emergence and this might serve as an aid to reducing weed competition at this crop stage.

In petri-dish studies, exudates of barley (cv. Thermi) have been shown to have an inhibitory effect on the germination of white mustard (*Sinapis alba*), indicating an allelopathic potential. Further studies of barley landraces and cultivars in Scandanavia for allelopathic ability against perennial ryegrass, showed that newer cultivars had reduced allelopathic capabilities but that barley cultivars inhibited the root growth of the ryegrass by between 42 and 70 per cent. On the other hand, barley cultivars were also susceptible, to a greater or lesser extent, to the allelochemicals produced by wild oats and some other weeds.

Pre-sowing harrowing and sowing in darkness, either at night or with covered implements, has significantly reduced weed numbers and weed biomass at early crop stages, although weeds still emerge later and grow unless the crop is able to make the most of the early advantage and smother the weeds.

SPRING BARLEY

Spring barley is more competitive than spring wheat or field beans against the perennial grasses, common couch and black bent. Barley cultivars differ in their competitive abilities against weeds and some studies of competition between spring barley cultivars and wild oat indicated that root competition alone accounted for observed yield reductions. However, plant height appears to be an important factor in decreasing weed biomass, as demonstrated in a series of pot experiments, which showed that taller cultivars reduced biomass more than shorter ones. In another study, there was a good correlation between maximum canopy height of the crop and a reduction in weed dry matter in a comparison of seven spring barley cultivars with weed weight up to 50 per cent less under some cultivars. In further studies, cultivars that developed good early ground cover suffered lower subsequent weed infestations. In summary, the use of different cultivars, each with differing dates of maturity, vegetative habit, height and competitive ability, offers good opportunities for weed suppression and, in this regard, it has been suggested that information on these characters, from variety trials based on conventional cropping, can be used by organic growers.

As a rule of thumb, relative time of crop and weed emergence are also major factors in determining the severity of loss to weeds; the greater the difference, the less the competition to the crop from weeds. In competition studies with spring barley, weeds allowed to grow for different periods after crop emergence, before being removed, had little effect on yield if removed by ten weeks from crop emergence. Where the crop was kept weed free for different periods, then left to become weedy, the weeds that emerged eight weeks after the crop had little effect on yield (reducing yield by 8 per cent). The main period for weed competition indicated by this study was therefore between eight and twelve weeks after crop emergence (although weed emergence in this experiment was limited by low rainfall).

Mechanical weed control may be necessary at times in spring barley. Research has shown that two harrowing operations from the two-leaf crop stage onwards, gave acceptable weed control, although some studies suggest that selectivity between the crop and certain weeds is too low at the two-leaf crop stage to be effective. Harrowing may also increase the competitive ability of surviving weeds. In field studies it has been possible to attain levels of weed control comparable to those achieved with herbicides using post-emergence harrowing alone, selective harrowing alone or

Fig 85 An infestation of couch in barley.

selective harrowing following earlier mechanical control. In one study, harrowing once pre-emergence and once early post-emergence, reduced weed biomass by 26 and 52 per cent in different years. The addition of a late post-emergence treatment reduced the weed biomass by 72 and 75 per cent on average. Weed control was more effective at 20cm (8in) row spacing than 12cm (5in). However, in practice, at low weed pressure, many of these studies have shown that harrowing is unlikely to increase crop yield significantly and increasing the numbers of passes raises the risk of crop injury for what becomes a decreasing level of improvement in weed control.

When using mechanical weed control, crop damage is possible. In a comparison of spring barley cultivars, many of the taller, high yielding cultivars with a high leaf-area index (LAI) tended to be less tolerant to post-emergence weed harrowing, compared with low yielding, shorter cultivars with a low LAI under the conditions of the trial (under dry conditions using an Einbock spring tine harrow at crop growth stages 13 and 22, to a tillage depth of 3cm/1in). In this case, and despite the damage, the tall cultivars remained the highest yielding. In the Netherlands, harrowing was found to cause a slight yield reduction because of the extra wheel tracks.

Where perennial grass is a problem it may be dealt with by control operations in barley stubble. One successful trial included one or two passes with a rotovator, the second when regrowth had one or two leaves, followed by ploughing, cultivation and drilling with spring barley. In this trial there was considerable foliar regrowth at the time of ploughing but no appreciable rhizome growth, the rotovator being set to cut at 12cm (5in). Shoot counts in spring showed a reduction of 80 to 90 per cent in common couch and black bent, the higher figure from two passes with the rotovator. The treatment did not result in complete eradication and would need to be repeated in subsequent years to maintain control.

Spring barley can also be used as a cleaning crop where rough meadow grass is a problem, as seedlings of this weed that emerge in spring barley do not become vernalized and are incapable of flowering. Provided that any over-wintered plants are destroyed by seedbed preparations, no fresh seed is produced. If any seed is shed, it is important to prevent this being incorporated into the soil where it may become dormant. Post-harvest cultivations should be delayed to allow the seed time to germinate.

MALTING BARLEY

Malting barley fetches a premium return, compared with fodder barley, and a prototype, modular, decision support system has been constructed to make the best use of mechanical weed control opportunities. Although experience in its use is limited, it could potentially be useful for those growing organic malting barley. Modules within the support system predict the effects of weeds and weed control, using data from a series of weed control trials conducted at three crop stages: pre-emergence harrowing to kill emerging weeds, post-emergence harrowing (at the two- to five-leaf stage), post-emergence harrowing at tillering (using the harrow tines inter-row). The decision support system determines the time to harrow as late as possible, close to crop emergence, based on day degrees. The benefits of post-emergence weeding are predicted on the basis of the amount of weed, the species present and their competitive ability.

WINTER BARLEY

Winter barley does not compete well against weeds, which when added to the limited opportunities for mechanical weed control in the autumn, make winter barley a difficult crop to grow organically. In a survey of winter barley crops in the autumn of 1988, the most common broad-leaved weeds were chickweed, field speedwell, the mayweeds and cleavers, amongst others. The most common grass was annual meadow grass followed by wild oat and black grass.

In a mechanical weed control trial in Spain, in winter barley grown at a conventional 12cm (5in) row spacing, pre- and post-emergence harrowing at speeds of 4km/h (2.5mph) tended to give better post-emergence weed control where speeds between 2 and 8km/h (1.2–5mph) were evaluated. Weeds were killed by a combination of soil burial and uprooting. A layer of soil 0.8–1.4cm (3/10–3/5 in) deep was thrown into the crop row. Soil texture had a major effect on soil movement but there was no consistent effect of tractor speed. Tap rooted weeds were better controlled when they were still small.

Weed Management in Maize

Maize is grown both for cobs (sweet corn) and as a cereal (fodder or silage) crop. Weed management in sweet corn is discussed in weed management in field vegetables below. As a fodder crop, maize has high nutrient and weed control requirements, together with a preference for drier conditions. Maize is very sensitive to weed competition at early stages and relatively short periods (two to four weeks) of weed competition, during early growth of the crop, may be sufficient to reduce final yield. In weed competition studies, maize has been left weedy for up to twelve days after emergence without loss of yield, and can be left to become weedy from day thirty onwards without loss. The crop therefore needs to be weed free in the intervening period. In Italy, in the field, the optimum time for weed removal to prevent significant yield loss was four to six weeks after crop emergence. In other studies, yield loss has been observed to increase with any delay in weeding but there is commonly no significant interference with crop growth when weeds were allowed to develop in the weed free crop from the seven- to eight-leaf stage.

The seed should be drilled into a clean, stale seedbed. Late-April or early-May is considered an appropriate time to sow maize in central or southern England but sowing later, in mid-May, allows the crop to get away more rapidly and out compete weeds. Sowing in narrower rows, 37.5cm (15in) rather than 75cm (30in), also gives a denser crop, which aids weed suppression.

The crop may be chain harrowed one week after drilling. In silage maize, in the Netherlands, harrowing early, that is between crop emergence and the two-leaf stage, did not affect yield if the crop was drilled deep enough (6cm/2.5in) and plant numbers were high. Harrowing four or five times, followed by hoeing to bury the weeds, also gave adequate control.

Once the rows appear, the crop is generally inter-row weeded at two-week intervals, until the crop is too tall, around the four- to six-leaf stage, by which time the crop canopy should hinder further weed growth. The last cultivation should aim to move soil into the crop row to bury intra-row seedling weeds. Some hand weeding may be needed within the rows. One study compared mechanical weeding of forage maize with a spring-tine duck-foot hoe weeder, a brush weeder, a ground driven rolling cultivator ('Lilliston') and a harrow tine weeder ('Tearaway') carried out at early, middle and late crop stages (three- to four-leaf, five- to six-leaf, six- to seven-leaf stages), with one to three passes. Early treatment increased maize yield but did not necessarily lead to a reduction in weed biomass. More than one pass gave better results than a single pass. The hoe and brush weeder were more effective than the tine weeder and rolling cultivator. In other studies, a combination of manual hoeing and mechanical weeding gave better results than manual hoeing or mechanical weeding alone, but might not be practical on a large field scale.

Maize can be flame weeded when it reaches a height of 10cm (4in) and can be flamed until canopy closure. The crop row is flamed across with burners mounted in pairs, but staggered to avoid overlap, and set at an angle of 30 to 60 degrees from the horizontal. If flamed earlier than this,

Fig 86 Maize, here grown for sweet corn, with weed infestation.

tall maize should not be treated again until it reaches a height of about 18cm (6–8in). In Italy, mechanical control and thermal weed control were compared with chemical control. Mechanical control consisted of inter-row cultivations at the four- to five-leaf stage of the crop and again at the ten- to twelve-leaf stage, when ridging was also carried out. Weed control was 25–44 per cent lower than with chemicals and yield was reduced by 6–18 per cent. The thermal treatment included inter-row tillage with band flaming in the row at the one- to two-leaf crop stage, inter-row tillage plus band flaming at the base of the crop stalk at the four- to five-leaf stage, and band flaming at the base of the crop stalk at the ten- to twelve-leaf stage, plus ridging. The thermal treatment gave good weed control but was likely to be more expensive than the use of herbicides.

In other studies, a surface mulch of dairy manure compost at 60t/ha (25t/ac) applied after drilling did not appear to inhibit weed growth, whilst a black polythene mulch laid after crop emergence, with holes cut to let the crop plants through, did reduce weed development. In the latter case, crop growth was also enhanced due to higher soil temperatures under the covers. Undersowing with a legume, once the maize is established, protects the soil from erosion when the crop is harvested. Intercropping maize with kale for silage production can have advantages, in terms of yield and weed suppression, over the crops grown alone.

Weed Management in Oats

Oats can do well in soils with low nutrient availability and are able to compete against weeds, although they should not be grown more than once in succession because of the risk of increased nematode populations. It has been suggested that selecting for a high rate of germination at low temperatures will improve the competitive ability of oats against weeds, particularly wild oats, during the early growth stages. Oats are reputed to have good allelopathic potential, especially older varieties, and are often regarded as a cleaning cereal crop by farmers.

Oats can be grown with peas and vetch as a high protein silage crop. It is useful as a clean-up crop because it smothers weeds. A sowing mixture by weight of 60 per cent oats, 24 per cent peas and 16 per cent vetch is considered a good mix. The crop should be sown in April or May and should be cut when the peas are formed but still soft.

SPRING OATS

In a study of competitive ability against weeds, there was little difference in weed biomass between cultivars of contrasting straw length. High seed rates and narrow row spacing increase crop yield and higher seed rates significantly reduce weed biomass. Altering row spacing had less effect but weed growth was greater at wide row spacings.

Post-emergent weed control is generally by harrowing. In one study, mechanical weeding with a 'Lely' weeder, carried out twice, at growth stages thirteen to fourteen and mid-tillering reduced weed biomass but did not increase crop yield. Harrowing depth is important in maintaining good weed control at later crop stages. In one study, at the three-leaf crop stage, harrowing at 2 or 4cm (4/5 or 1 3/5 in) depth gave a 40 and 60 per cent reduction in weed biomass, respectively. Late harrowing at the same depths, at the six-leaf crop stage, gave a 30 and 40 per cent reduction in weed biomass.

It has been shown in other work that the crop can be severely affected by soil covering when spring oats are harrowed at the one-leaf stage, although this is generally too early to harrow as few weeds have emerged. Both weed control and crop yield are better following harrowing at the two- or three-leaf crop stage. In spring oats at the three- to four-leaf stage, harrowing across the crop rows resulted in greater soil covering of the crop than harrowing along the rows; increasing the driving speed also caused greater soil covering. Although the soil covering had no apparent effect on crop growth, at the higher tractor speeds of 9 and 33km/h (5.5 and 20mph), crop yield was adversely affected in two of the three years of testing. Good weed control was obtained by harrowing at 5km/h (3mph) but the direction of harrowing did not matter.

Attention to detail in harvesting can also be important. A preliminary study in Sweden demonstrated that the number of weed seeds left on the ground after combining oats was much higher than when the crop was harvested with a binder, dried in shocks and then threshed. There were over fifteen times more fat hen and cleavers seeds and 175 times more creeping thistle seed after combining, but bind harvesting is unlikely to be a popular choice!

WINTER OATS

There may be limited opportunities for mechanical weed control in the autumn but this will depend on weather conditions.

Weed Management in Rye

Rye is able to compete against weeds, has low nutrient requirements and will grow on poorer, lighter soils. However, there may be limited opportunities for mechanical weed control in the autumn because of tillering. On the other hand, rapid spring growth and long straw length can give good weed suppression. The mechanical weed control methods employed in other cereal crops should give effective control of weeds in rye.

Weed Management in Triticale

Triticale, a cross between wheat and rye, is a popular choice of cereal for organic farmers. It can also be used as a forage crop or green manure. It is generally reckoned to be competitive against weeds. Once again there may be limited opportunities for mechanical weed control in the autumn, otherwise similar practices to wheat (*see* below) and barley (*see* above) should be effective in controlling weeds in this crop.

Weed Management in Wheat

Wheat is often the first cereal in a rotation after a grass/legume break or after a break crop such as potatoes or maize, although it should not be grown more than twice in succession due to increased weed competition, disease risk and the decline in soil nitrogen.

Modern cultivars with a short upright habit are not suited for good weed suppression. It is sometimes thought that spring varieties are more competitive against weeds but a late sown well-established winter crop can also give good suppression of spring emerging weeds. Older varieties, released between 1880 and 1950, are said to suppress weeds more than the current modern varieties, although studies have shown that recent introductions can also perform well. The characteristics that improved competitive ability against weeds were: high tiller numbers, greater plant dry weight and increased seedling growth rate, but the most important was plant height. Long strawed cultivars can help with weed suppression but high crop aggressivity against weeds is most associated with a high crop LAI. Wheat roots produce exudates that have been shown to inhibit the growth of charlock and bioassays have shown that wheat cultivars differ in their ability to inhibit charlock growth, due to differences in the content of allelopathic chemicals.

While choice of cultivar can aid weed suppression, seed rate is probably more important. Optimization of row spacing and seed rate may not affect early weed density but can contribute to reducing the growth of weeds. Both factors are also important in determining crop yield. Mathematical models suggest that sowing wheat in square, rhomboidal and rectangular patterns should increase yields, compared with sowing in traditional rows. However, field trials in Australia have shown no consistent advantage at any particular sowing pattern. Likewise, it has been noted by some workers that the shading effect of tall cultivars increases when the crop is sown in an east–west direction, but this will depend on latitude, as well as many other factors like varieties used.

SPRING WHEAT

Spring wheat is less competitive than winter wheat and can have greater weed problems. Delayed seeding of spring wheat has been used to allow wild oat to be controlled, but any delay in sowing dates also reduces the potential yield of the crop. Some trials indicate small differences in weed suppressive abilities between cultivars. For instance, the cultivar Baldus was found to be more competitive against charlock than Canon, due to differences in canopy structure

Increasing sowing density and changing sowing pattern have improved weed suppression in spring wheat. For instance, increasing sowing density from 400 to 800 seeds/m^2 (40–80/ft^2) and changing the spatial pattern from 12.8cm (5in) rows to a uniform grid, reduced weed biomass by 60 per cent and increased yield by 60 per cent, with an average weed biomass 30 per cent lower where the crop was sown using a uniform pattern. In a trial with spring wheat sown at two seed rates, 140 and 180kg/ha (57 and 73kg/ac), and at three row spacings, 10, 20 and 30cm (4, 8 and 12in), weed biomass and dry weight was reduced as row spacing decreased and crop sowing density increased. Reduced weed seed production followed the same trend. In studies of crop density and spatial arrangement using normal rows, uniform and random patterns at crop densities of 204, 449 and 721 seeds/m^2 (20, 44 and 72 seeds/ft^2), weed biomass was lower and wheat biomass higher in random and uniform patterns than in normal rows. Increased crop density decreased weed biomass in all three spatial arrangements. There is some indication that intra-specific competition decreases crop yield at a sowing density of 1,000 seeds/m^2 (100/ft^2) included in the later trial.

At early crop stages, mechanical weed control is possible using a harrow comb weeder. At later crop stages, tine weeders are more likely to be more effective and to do less damage to the crop. The number of passes depends on weed and crop density, but increased numbers of passes give diminishing returns in terms of weed control and carry a higher risk of crop damage. Sowing arrangement of the crop can also influence weeding effectiveness, although different sowing rates and row spacings do not have a consistent effect on yield, and results can be very situation specific. There is a trend towards drilling crops in rows, which can allow more aggressive weeding with less crop damage (*see* Winter Wheat below). Mechanical weeding in June and July (in addition to harrowing) with a spring tine harrow in narrower crop rows has been shown to give a better result than inter-row weeding with a 'Rabe' hoe with V-blades at wider 30cm (12in) rowspacing.

A number of simulation models predicting the effects of weeds on the crop and the efficacy of weeding operations have been developed but many are not particularly user friendly, or are mainly directed at herbicide use. For instance, a simulation model has been developed to evaluate the practical use of relative leaf area to predict yield loss from weeds at an early crop stage and the probability that control strategies will be economically justified. However, crop prices can vary from year to year and throughout the year, and this, coupled to changes in the cost of control measures, obviously affects the predictions of the model and might limit its usefulness in practical farm situations.

WINTER WHEAT

Research has indicated that the critical period for weed control, which minimizes yield losses, spans from November to March. This contrasts with the usual practice, in the UK, of weeding in the spring, when it is operationally easier to do so. In field experiments in the UK to determine the critical weeding time in winter wheat, there was no time when weed control could be relaxed without some loss of yield. If a 10 per cent loss in yield was acceptable, then the crop needed to be kept weed free from November to March. If a 20 per cent loss was tolerable, then removal of weeds between December and February appeared to suffice. In competition studies with black grass and with chickweed, it was found that weeding could be deferred until mid-March without significant loss of yield. In contrast, competition studies in Belarus indicated that it was sufficient to weed by the full tillering stage in spring, although ideally winter wheat needed to be free from weeds as early as possible.

A survey of winter wheat (in conventional production) in autumn 1988, to assess the occurrence of broad-leaved and grass weeds, indicated that the most frequent broad-leaved weed was cleavers and the most frequent grass weed was common couch (found in 42 per cent of cereal fields). In other studies in organic sytems, the main weeds present were black grass, scentless mayweed, common chickweed and parsley piert. It was suggested that cultivars that developed good, early ground cover, suffered lower subsequent weed infestations and organic growers could use information on this character from NIAB variety trials based on conventional cropping. Some of the newer feed cultivars can also perform well when drilled in November and December, but seed rates may need to be increased to take account of the cold, wet conditions. Previous cropping history can also affect weed incidence. In one study, weed seedling emergence in winter wheat was greater on land where the previous winter cereal had been sown early rather than late. Seedling numbers of some weed species, such as black grass and red deadnettle, were reduced when sheep had grazed the previous cereal crop, but others, such as chickweed were unaffected. The cultivar used in the previous cereal could also affect seedling numbers in the following crop and there were significantly more charlock seedlings on land previously drilled with cv. Maris Widgeon than with either Genesis or Hereward.

Fig 87 Poppies in wheat.

Cultivar choice has been shown to have different effects in different years (and differences are often not statistically significant over a run of trials). Studies have shown that the choice of a tall cultivar (and increased nitrogen levels) can significantly reduce weed dry weight, weed density and weed diversity, due to lower light levels beneath the canopy suppressing weed growth. In Greece, a comparison of the competitive ability of a semi-dwarf and a tall cultivar against weeds, confirmed that plant height was important, but vigorous early growth and high tiller number also contributed to weed suppression. In the UK, many detailed studies have been made on cereal–weed interactions and comparisons of the ability of winter wheat cultivars to compete against weeds showed straw length to be an important factor in reducing weed biomass, although neither straw length nor weed biomass was always related to grain yield. It has also been noted that tall cultivars may actually favour the establishment of some weeds, for example, ivy-leaved speedwell.

Crop density and sowing pattern are also important factors in suppressing weed growth and limiting weed seed production. Weed biomass at harvest is generally reduced by higher seed rates, although seeding rate tends to make little difference to crop yield. Seed rates above 300 seeds/m^2 (30 seeds/ft^2) tend to have little extra effect on weed suppression, although weed seed production can be halved as crop density is doubled up to 200 crop plants/m^2 (20/ft^2). Wheat sown at narrower 12cm (5in) spacing can give greater weed suppression than that sown at the standard 18cm (7in) spacing. There is also a tendency to sow cereals in rows, as this can greatly facilitate inter-row weeding (*see* below). In any case, drilling after mid-October is better for reducing weed competitiveness; weed dry matter has been shown to be 20–30 per cent lower in late-sown winter wheat ,as compared to early sown crops, although yield can be reduced by up to 10 per cent.

Winter wheat can be established under vary variable situations. The first cultivation may be as early as July after a hay or silage cut or as late as September and, depending on circumstances, a range of cultivations might be performed to control (weed) regrowth and newly germinated weeds. Shallow ploughing may be required in mid-September to bury the trash. Where the crop follows a ley, and this is broken up by a rotovator, the first pass is the most important, as subsequent passes will mainly move around, rather than break up, the turf pieces. Regenerating docks may be a problem following a ley. The cultivations needed for final seedbed preparation will depend on whether the land is light or heavy in texture. From the point of view of utilizing the available nitrogen in the soil, the cereal should be sown as early as possible, although this conflicts with the advice to delay sowing to control weeds (*see* above). On the other hand, some studies suggest that weeds have less effect on the crop at low nitrogen levels and this might be advantageous in organic farm systems. For example, in one set of experiments with winter wheat given different levels of nitrogen, there was no weed competition at low levels of fertility and crop yield was reduced only when soil fertility was high and the weeds grew large. Other research suggests that nitrogen applications in winter wheat and spring barley do not increase weed numbers or total weed biomass, as greater crop growth reduces the light reaching the weeds, although there may be a differential effect on the biomass of individual weed species depending upon their exact nitrogen requirements.

Mechanical weed control operations are commonly carried out in winter wheat. The level of weed control achievable in a limited number of weeding operations will depend on the implement used and the weed species present. Mechanical weeding is broadly split between spring tine weeding and inter-row hoeing, although harrow comb weeders are often employed when the crop is small. However, results from a range of studies have indicated that removing the weeds from organic winter wheat crops only occasionally gives a yield benefit, with weeds only being a significant problem where winter kill thins the crop (seedlings).

Blind harrowing. At early crop stages, mechanical weed control is possible using a harrow comb weeder. The number of passes depends on weed and crop density. Blind harrowing with spring tines at four to five days after sowing is often followed by harrowing at the four- to five-leaf stage of the crop (at which point clover seed can be incorporated in undersown cereals). In winter wheat, in the Netherlands, harrowing was found to cause a slight yield reduction because of the extra wheel tracks. In this study, two harrowings, from the two-leaf crop stage onwards, could give acceptable weed control, but success depended on the time of sowing, how early the first harrowing could be performed in spring and on the weed species present. Early sowing or late harrowing resulted in poor control of black grass and scented mayweed. For cleavers control, harrowing with a tine weeder at an early crop stage can give a 79 per cent reduction in density of the weed. A second harrowing at a later crop stage improves the level of control. The cleavers seedlings need to be large enough to be caught by the tines. The harrowing does not reduce crop yield but a yield benefit may be found only when the weed is at a high density. A Swedish study has shown that late harrowing did improve yield in winter wheat (but not in spring cereals), whilst a harrow tine weeding ('Tearaway') carried out in June reduced weed biomass in some years but not others.

Inter-row hoeing is less sensitive to weed stage and so weeding operations can be postponed to spring when the crop is less vulnerable. Intra-row weeds are only partially controlled, depending on the amount of soil coverage in the row. In a study, hoeing in early April and again in late April gave better weed control than a single weeding. The early pass is important, as hoeing late may control the weeds but does not prevent weed competition. Increased driving speed does not improve weed control greatly. Inter-row steerage hoes based on computer vision have been developed to address the problem of accurate guidance between the crop rows to avoid crop damage and are now quite sophisticated. In one study, an inter-row steerage hoe, guided by computer vision, was compared with an 'Einböck' finger tine harrow weeder on different row widths, 12.5, 15, 18 and 22cm (5, 6, 7 and 9in), of winter wheat. Both the steerage hoe and the harrow reduced weed weights but there was no yield benefit. Row width did not affect yield either. However, the hoeing could be carried out over a wider range of crop and weed stages than the harrowing. Other results suggest that a row width of 16cm (6in) allows better weed suppression than wider or narrower row widths. If inter-row hoeing is unlikely to be attempted then rows as narrow as 7.5cm (3in) may be better to out-compete and smother the weeds.

In Denmark, winter wheat was grown on a 24cm (9.5in) row spacing to allow inter-row hoeing with a 16cm (6in) blade without altering the seed rate (so that within row density was higher). Post-emergence harrowing treatments with a flat blade tine weeder were applied immediately after hoeing, and early and late post-emergence harrowing treatments were made at speeds of 2–8km/h (1.2–5mph) in early April and two weeks later. Weeds were killed by a combination of soil burial and uprooting. A layer of soil 0.8–1.4 cm (3/10–3/5 in) deep was thrown into the crop row. More soil movement occurred at higher tractor speeds and when soil was dry. Tap rooted weeds were better controlled at early growth stages.

The best method in any situation will depend on the specific circumstances and be largely a matter of experience or preference. In one comparison of spring tine weeders and inter-row hoeing in winter wheat, efficacy of the spring tine weeder depended on timing and the species present. Tap rooted weeds were best treated in autumn and fibrous rooted ones in spring. On the other hand, the inter-row hoe was less time and species-dependent. In separate trials with a chain harrow (with shuttle shaped and round links), a rigid tine weeder, a spring tooth weeder and a hoeing machine (with duckfoot blades), the machines differed in their effects on different weeds. With the weeders and harrows, weeds at less than the three-leaf stage were reduced by 60–85 per cent by one pass. Larger weeds, with three or more leaves, were reduced by 15–30 per cent on clay and 33–63 per cent on loam. With the tine weeder, stronger tines, 0.8cm as opposed to 0.6cm (3/10 vs 1/5 in), gave better results on the clay soil. Adjusting the penetration angle and spring tension on the tines also helped. Two passes in opposite directions gave higher weed reductions with the weeders and chain harrow. The hoeing machine gave a 90 per cent reduction of the larger weeds, even on the heavy soil, but was restricted to the inter-row area and left the intra-row weeds. Nitrogen mineralization of the soil was increased by the more intensive disturbance due to hoeing and spring tooth weeding but not by the other machines.

A range of more novel weed management strategies have been developed or suggested from time to time, and some of these are briefly described below:

Inter-cropping. Studies of inter-cropped field beans and wheat, where the beans were sown first and the wheat sown a day later at right angles to the beans, have shown that the weed biomass under the inter-crop was lower than under the crops grown separately. The optimum crop density was 75 per cent of the recommended sole cropping density for each component crop. It was necessary to match the harvest dates of the crops by choosing an early maturing bean and a late maturing wheat, so that it was possible to harvest both crops together by combine harvester. The seed can be separated mechanically or left as a mixture for stock feed. Higher light interception by the crop canopy of inter-crops, resulting in greater competition for light, appears to be an important factor in limiting weed growth.

In another inter-crop system a method for direct drilling winter wheat into a clover sward has been developed. A pure stand of white clover is sown in spring, established through the summer, before being cut and ensiled in autumn. Winter wheat is then direct drilled into the clover sward. The cereal may be harvested for whole crop silage or taken through to grain harvest. The clover is allowed to regenerate before being grazed by sheep in the autumn and re-drilled with a second crop of wheat. The initial clover crop can be undersown in spring barley and the crops cut as whole crop silage. Spring cereals cannot compete with the clover once it is established, neither can annual broad-leaved weeds. Grass weeds, such as rough meadow grass, can become a problem and invasion by grass weeds caused yield reductions in some trials where herbicides were not used prior to drilling the cereal.

Mulching. A series of applications of mulch 3cm (1⅕ in) deep has been used to increase crop establishment and yield, and reduce weed emergence, but not where the compost was incorporated.

It is often difficult to unravel the various complex factors in a weed management strategy. One study that attempted to do this for winter wheat by comparing the effect of sowing date, stale seedbed, row width and mechanical weed control, found that treatments that improved yield when weed numbers were high, often led to a yield reduction when weed pressure was low. Weed numbers and biomass were less when the crop was sown in mid-October compared with late-September but the potential yield was also less. Potential yield was also likely to be lower with wider row spacing but this allowed more intensive mechanical weeding to be used to deal with dense weed stands. The false seedbed reduced weeds in the crop but was likely to lead to a delayed drilling date and hence a reduced yield potential. The effect of mechanical weeding depended on the weed pressure. At low weed pressure, mechanical weeding was likely to cause a decrease in yield. At high pressure, mechanical weeding gave a yield increase compared with the untreated crop.

Given this complexity there have been many attempts to develop weed management support systems in order to provide a tool for the effective transfer of optimum advice on weed control in winter wheat to farmers. As far as the weeds are concerned, such a system would predict likely yield losses from the weeds present, the impact of weeding time on efficacy of control and on crop yield, the effect of the crop on the weeds, the potential seed production by any weeds left to survive and longevity of those seeds in soil. Weeds species would need to be ranked on the basis of their competitive ability and then grouped into those that it is critical to control, those that should be controlled and those that need not be controlled. The system would also provide an assessment of the impact of rotation, choice of primary tillage, drilling date and nature of post-harvest cultivations. As of yet, there are no practical models in place, and farmers are best advised to keep records, build up their experience and take as many learning opportunities as possible in order to improve weed management practice. Some models have been developed that apply to specific situations. For instance, a model that has been developed for the population dynamics of barren brome in winter wheat suggests that crop management, particularly the form of cultivation, is more important than weather conditions for weed development. Controlling weeds may also be important to prevent weed seeds returning to the soil and increasing problems in subsequent crops.

Weed Management in Field Vegetables

Field vegetables cover a wide range of crops. Organic field vegetables are often grown in intensive production systems and the value of the crop can be high, compared to those in arable or other more extensive systems. This in turn means that weeds can cause considerable economic damage. In these circumstances, weeding costs can be high but economically viable. Below, we describe some key features of the more common crop types and how they bear on weed management decisions.

Weed control in field vegetables is helped by an appropriate rotation that avoids the build up of large weed populations. Making the most of weed control opportunities in one crop can frequently ease weed control problems in the crop that follows. The preparation of false or stale seedbeds is also indispensable for reducing weed numbers, especially in the direct sown crops, but will also be valuable in transplanted crops. Secondary cultivations may be used to kill emerged and germinating weed seedlings but must be shallow enough to avoid stimulating a further flush of weed emergence. Flame weeding will also kill the emerged seedlings and without disturbing the soil. Careful timing of flame weeding is needed to kill the maximum number of weeds without damaging the emerging crop seedlings.

Fleece covers are sometimes used for early production and for pest control in some vegetable crops (especially brassicas and salads). Unfortunately weeds also benefit from the conditions under the covers, emerging in greater numbers and growing faster, so that regular inspection is needed to ensure weeding windows are not missed. Efficient systems for removal and replacement of fleece will help in these situations and, in the case of pest control, care should be taken not to remove the fleece when the pest is active and likely to find the crop; covers should be kept off for a short a time as is practical.

Weed Management in Allium Crops

Alliums (onions, leeks, garlic) tend to germinate and grow slowly, they generally have rather few leaves and, for most of the growing season, the soil is not fully covered. Weeds can therefore germinate over a long period. Allium crops are also very sensitive to weed competition, so weed control is of particular importance. Weeds are more of a problem in crops grown from seed than in transplanted or vegetatively propagated crops. In general, a drilled onion crop will produce no yield if it is left unweeded. At crop harvest, weeds foul undercutting and lifting machinery, and prevent onion bulbs drying in the windrow.

The basic strategy for weed control in allium crops starts with the choice of field; rotation should also be used for disease control. The structure of the soil is important, in the sense that the preparation of the seedbed is a crucial part of any weed control strategy. In the case of alliums,

Fig 88 Onions infested with Persicaria *spp.*

the seedbed should be fine, crumbly and level. It should be borne in mind that volunteer weeds can be a problem when following crops such as potatoes, cereals and oil-seed rape. Allium crops should not be sown or planted too early, otherwise the crop will grow slowly, giving the weeds the advantage of a longer germination period. In organic systems, bare root or module transplants can help give the crop an advantage over weeds, especially when combined with stale seedbed preparation. Because perennial weeds are very difficult to control in allium crops, they have to be controlled in the preceding crop(s).

Crop establishment generally follows normal primary and secondary cultivation operations. The main methods of weed control are mechanical and thermal. Mechanical control includes harrowing and hoeing, while thermal control involves flame weeding to control small seedling weeds. The success of these methods depends on timing, on weather and soil conditions, and on the composition and density of the weed population. Some alliums, such as early bulb onions, have sufficient value to be grown through sheet mulches, normally black plastic or (biodegradable) starch polymers that are effective in preventing weeds emerging over the entire cropping bed.

Weed Management in Beets (Spinach Beets and Chards)

Apart from fodder and sugar crops, beet crops are also grown as a root vegetable (beetroot) or leaf salads. Weed control in leafy salad crops is described below. Land used to grow beetroot should be free of perennial weeds and the normal, traditional primary and secondary cultivations are used to establish the crop in a weed free bed. False or stale seedbeds will reduce weed numbers in the growing crop. The optimum time for weed removal is around three to four weeks after crop emergence. Once the weeds have been removed, the crop has some capacity for recovery from a check to growth due to the weeds. As with sugar beet (*see* above), once the crop has emerged, regular inter-row cultivations with brush weeders, ridgers, steerage hoes, finger tines or similar, will deal with weeds between the rows. There will be some effect on intra-row weeds and hand weeding may occasionally be required to avoid damage to the roots.

Weed Management in Brassica Crops

In organic rotations, brassicas are generally placed at the beginning of a rotation and can follow on directly after a ley or a cereal crop. The land should be ploughed and prepared early in the year to allow time for cultivation before planting that will kill any emerged weeds. Following crop establishment, weeds should be controlled without delay. Crop cultivars vary in the ability to suppress weeds because of differences in morphology.

In vegetable brassicas, the choice of cultivar may be limited by the need to schedule maturity to achieve continuity of production. The use of crop covers to advance maturity and protect vegetable brassicas from insect pests will hinder weed control and may enhance weed emergence and growth.

Brassicas are also often transplanted and this can help to give the crop a head start over any weeds, especially if a stale seedbed is prepared. Many vegetable crops, including brassicas, are grown on the bed system and cultivators and other implements should be matched to the bed widths. Precise row spacing, and careful alignment of cultivating tools, facilitate mechanical weed control. Weeding is repeated as necessary, and cultivations can become more thorough as the crop develops and becomes firmly rooted. When the crop plants are large enough not to be buried, tools can be arranged to move a 2.5 cm (1in) layer of soil towards, and into, the crop row. Where mechanical control is not possible, hand weeding may be required.

BROCCOLI

Transplanted broccoli rapidly develops a broad, shading leaf canopy. Inter-row weed control, with a row crop cultivator, spider gang tool or a

Fig 89 Weeds, principally chickweed, left in cabbage rows after inter-row weeding.

brush hoe, fifteen to twenty-five days after planting, has given good weed control. Flexi-tine harrows provide adequate, within and between row, weed control but damage poorly established crop plants and reduce yield. Mechanical weed control with round and flat flexi-tines, brush hoe, rolling cultivator or shovel cultivator, alone or in sequence with flexi-tine weeders just prior to weed emergence at thirteen to twenty-six days after crop planting, have been shown to control weeds. Finger weeders have also been observed to give good intra-row control of small weeds. As with cabbage, the broccoli crop has some tolerance to flame weeding, and treatment at two weeks after crop planting has been evaluated, but some care may be needed in practice!

BRUSSELS SPROUTS

These are generally a weed suppressing crop and can be ridged up to improve weed control around the sprout plant base. The wide spacing of the crop gives ample scope for mechanical weed control. Transplanted Brussels sprouts have an initial advantage over emerging weed seedlings, nevertheless, if left unweeded, sprout yield is likely to be reduced by 13–24 per cent. Close crop spacing (45 x 45cm/18 x 18in) increases weed suppression, compared with wider spacing (61 x 61cm/24 x 24in) but within crop competition then tends to reduce the yield of individual sprout plants. At closer row spacing, access for tractor steerage hoes becomes more difficult after early August, as the plants grow in stature. Early weed control is therefore vital to avoid yield loss. As with cabbage, the Brussels sprout has some tolerance to flame weeding, and treatment at three weeks after crop planting has been tried successfully; but once again care should be taken.

CABBAGE

Different types of cabbage include spring, summer, autumn and winter cabbage like winter white and Savoy types. Obviously the details of weed control will differ between them. Good weed control is essential to maximize crop uniformity, quality and yield. Crops are normally established following conventional primary and secondary cultivations. Cabbage may be direct sown (e.g. some spring greens) but is more likely to be grown from bare-root or module raised transplants. The key period for weed control is the first four weeks after transplanting. Weeds that emerge after this are less likely to compete with a well established crop. In the drilled crop, a single weeding at three weeks after crop emergence, can give yields similar to those of the weed free crop. In transplanted cabbage, a single but thorough weeding, three to eight weeks after planting, has been sufficient to prevent yield loss.

Plant spacing can be adjusted to some extent to allow earlier ground cover within the row to smother weeds but the main method of weed control will be mechanical, by surface cultivations, using tractor mounted cultivators between the rows. Experience has shown that brush weeders, finger tine weeders, harrows and tractor drawn hoes can all be set up to work well, but that the number of passes needed will depend on the weed population. As the crop develops, the leaves spread into the inter-row and care is needed to minimize crop damage.

Flexi tine weeders with either flat or round tines, used fourteen and twenty-four days after transplanting, have given adequate weed control and only slightly damage the cabbage. Similar results were given by flex-itime cultivations at ten and twenty days, followed by an S-tine row crop cultivator at thirty days after planting. The brush weeder has more flexibility to work in wetter conditions than other implements and works to flick small weeds out of the soil. Brush weeders and tractor drawn hoes are available as steerage hoes requiring a second operator to guide the hoes or brushes between rows, the main advantages being the increased work rate, which is about three times faster than when the tractor driver has to guide the hoe, and the ability to steer very close to the row, which gives some intra-row weed control as soil is moved into the crop row.

Fig 90 Clean cabbage crop after mechanical weeding.

Front mounted hoes give the driver a better vantage point and may help improve weeding performance with only one operator.

Flexi-tine weeders can have some intra-row weeding effect (although they may damage the crop) but the best effect may be achieved by use of finger weeders mounted on inter-row hoes or dedicated finger weeders, as long as the cabbage is firmly rooted. In both operations, soil should be drawn up towards the plants to bury weeds in the row. Hand hoeing between the rows may be used during early crop establishment, but the crop rapidly forms a dense leaf canopy that helps to suppress further weed development. Hand labour requirements for weeding have been estimated at fifty person hours per hectare (twenty person hours per acre).

Transplanted cabbage has a relatively high tolerance to heat, enabling band-flaming to be used along the crop row. However, the level of tolerance within each individual crop will depend on the growth stage of the crop and the waxiness of the leaves. After crop establishment, thermal weed control may be tolerated by cabbage, if shields are used and the burners are directed away from the crop row. The aim is to kill seedling weeds in the inter-row with minimal soil disturbance. At an early stage, most broad-leaved weed seedlings will be killed but grass weeds have a basal growing point and therefore a higher tolerance of flame weeding. Perennial weeds are also unlikely to suffer any long-term damage. The tolerance of most weeds increases with age and growth stage. Flame weeding is not likely to be cost effective when compared with mechanical inter-row weeding.

Mulches laid directly on to the soil surface provide a physical barrier to weeds. Black polyethylene is generally left down for the duration of a crop but studies have been made where the sheeting has been laid on the seedbed for much shorter periods and then lifted before planting brassicas. The short-term covering of the soil with black polyethylene reduces subsequent weed emergence, giving the crop an advantage over the weeds. Straw mulches have been evaluated in trials but there appears to be a detrimental effect on yield, probably due to a temporary shortage of nitrogen as the straw decomposes.

CAULIFLOWER

Cauliflower should be placed early in the rotation to take advantage of higher fertility after grass–clover leys. The seedbed should be weed free at the time of planting. One or two inter-row passes with a brush weeder, steerage hoe or finger weeder should be sufficient, unless weed pressure or weather conditions result in poor control. Crop canopy closure should suppress later weed emergence and growth. As with cabbage, the cauliflower crop has some tolerance to flame weeding and treatment at two weeks after crop planting has been successful.

KALE

In transplanted kale, a single cultivation ten days after planting was sufficient to prevent yield losses due to weeds. A further cultivation ten days later reduced weed biomass but did not improve crop yield. Inter-seeding with the winter cover-crops, winter rye, hairy vetch (*Vicia villosa*) or a mixture of the two, after the final cultivation, allowed the cover crop sufficient time to establish without sacrificing crop yield. Inter-cropping maize with kale for silage production can have advantages, in terms of yield and weed suppression, over the crops grown alone.

SWEDE

The crop is often sown late to avoid turnip fly and mildew, leaving time for cleaning the land before sowing. Inter-row cultivations can be used after crop emergence. The optimum time for weed removal is six weeks after sowing or two to four weeks after crop emergence. Weeds that emerge and compete with the swedes for up to twenty-eight days after crop sowing, have no effect if the weeds are removed and the crop then kept weeded. Swedes have some ability to recover after removal of weed competition.

TURNIP

Turnip has been considered a cleaning crop in the rotation because regular inter-row cultivations can be carried out until canopy closure. Sowing time is generally mid- to end of May but wet conditions may delay this until June. The crop is often sown late to avoid turnip fly and mildew, which allows time for seedbed preparation and removing weeds before sowing. Inter-row cultivations can be used after crop emergence. By sowing turnips after a ley, all the weeds that germinated could be destroyed by cultivations. The optimum time for weed removal is two to four weeks after crop emergence. In the past, the inter-rows were horse hoed and the rows were hoed when the crop was singled. A second weeding was carried out later, if required.

Weed Management in Lettuce and Salads

There is a low or zero tolerance for weed contamination in salad crops because of the recent market trend for salad packs. In addition, weeds can have severe repercussions on yield. In drilled lettuce left weedy through to harvest, yield losses can be 100 per cent, often due to the detrimental effect on crop quality, which reduces marketable yield to zero.

Any site used for salad production should also be free of perennial weeds. The seedbed should have a fine tilth and, if prepared early enough, provide an opportunity for a weed strike and a final shallow cultivation to reduce the weed population before planting. This stale seedbed technique can be effective, if there is sufficient moisture for weed germination. Flame weeding applied post weed emergence, and before lettuce planting, can be successful when weeds are still small. If fleece is used, regular inspections should be made under the fleece to check how weeds are developing in comparison to the crop.

Precise row spacing and careful alignment of cultivating tools facilitate mechanical weed control. Closer spacing of lettuce may suppress weeds better but it will make mechanical or even hand weeding difficult, should it become necessary. A minimum row spacing of 25cm (10in) is needed to allow mechanical weeding. One or more passes with a brush weeder, steerage hoe or tine weeder may be required, depending on the weed population. All can be equally effective but certain implements are better than others in particular conditions. A first pass may be needed three to four weeks after planting. Weeds should be dealt with while small. Hand labour is not normally needed but has been estimated at twenty to twenty-five person hours per hectare (about nine person hours per acre). In comparison, tractor hoeing is likely to take 5–6h/ha (about 2h/a).

Weed Management in Solanaceous Crops

The Solanaceae are a large plant family including crops such as aubergines, chilli peppers, tomatoes, potatoes and sweet peppers. Most, except potato, are generally grown under protection in commercial situations in the UK. Potatoes are generally grown on a field scale and are discussed in more detail below. For protected crops, soil preparation is important to prevent weeds establishing and building up. Generally, hand weeding is the only viable and practical method of weeding under protection, although undersowing tall crops, such as tomatoes, with a low growing legume, such as trefoil, is a possibility, which can also help with fertility building.

The necessity for weed control operations can be minimized with a targeted watering regime that directs water at the base of the crop plants and keeps most of the soil surface dry, preventing weed seed germination and establishment. In some cases, woven plastic mulches or sheet mulches (black plastic or biodegradable starch plastic) are used to control weeds under protection, although thought needs to be given to watering the crop and preventing overheating, especially in the latter case.

Weed Management in Squashes (Marrows and Courgettes)

There is usually the opportunity to prepare a stale seedbed and then use inter-row cultivations and hand weeding to deal with subsequent weeds in squashes. Cross-hoeing has been practised in widely spaced pumpkins and squashes. Marrows and courgettes are also often transplanted into black polyethylene mulch or, more appropriately, into biodegradable starch mulches. The system is expensive due to the costs of materials and labour but can also help to advance the crop, as well as controlling weeds. The plastic mulch can be combined with the use of crop covers for earlier harvesting.

Weed Management in Peas and Beans

Legumes like peas, broad beans, French beans and runner beans can be grown on a full field scale or as part of an intensive vegetable production rotation. In the latter case, they will normally be grown in a bed system. Weeding is generally difficult and restricted to hand hoeing, especially where stakes are used, as in runner beans or climbing French beans. Good seedbed preparation is therefore necessary and stale seedbeds may help if adequate time is available to allow weeds to germinate and be controlled with secondary cultivations or flaming. Once again, it may be possible to undersow the climbing beans with a low growing legume that suppresses weed growth and contributes to fertility building. More details on weed effects and timing of weeding are given below in the section on weed management in legumes.

Weed Management in Perennial Vegetables

There are a few crops that are generally considered as perennial vegetables, including asparagus and Jerusalem artichokes. It is best to grow perennial vegetables in areas free from perennial weeds, and to try and prevent their encroachment into the crop.

ASPARAGUS

It is vital that the land is free from perennial weeds before the crop is established. Cultivation is not possible within the row during spear production. Soil cultivation to build up and maintain the planting ridges, allows for some weed control. Ridging may be done by hand but mechanization is used on larger plantings. In early autumn, the fern is cut down and cleared and the soil on the ridges and in the valley bottoms lightly cultivated to improve the tilth for ridging in late autumn. Shallow cultivations may be possible to control winter annual weeds that have emerged. In spring, cultivations are kept to a minimum on the ridges to avoid spear damage. The in-row weeds may be controlled by hand weeding, mulching or with the use of geese. Flame weeding has been used against annual broad-leaved weeds but little information is available on crop stage. A low growing living mulch that will suppress weeds is another possibility but there may be some competition with the crop.

JERUSALEM ARTICHOKES

Land should be cleared of perennial weeds in previous rotations. The crop is fast growing and able to smother annual weeds. The artichokes themselves are liable to become volunteer weeds in following crops if not all the tubers are harvested. The plant is capable of rapid spread by means of rhizomes and tubers, although tubers and rhizomes that fail to shoot do not persist in the soil for more than a year. In some cases it might be better to treat this crop as an annual to avoid problems with volunteer plants, in which case pigs can be used to clean up the ground and remove all tubers and rhizomes.

Weed Management in Radish

In favourable conditions, radish emerges earlier and grows faster than naturally occurring weeds. The crop is able to mature before the onset of weed competition. In addition, few weeds have reached the flowering stage before crop harvest, so seed shed is not a problem. There is therefore no need for a carefully timed weeding, if crop yield is the only concern.

Weed Management in Sweet Corn

Weed competition trials with rows 25, 51 or 89cm (10, 20 or 35in) apart, with plants 30cm (12in) apart in the row, demonstrated that in unweeded plots there was somewhat less yield loss in the narrower crop rows. Keeping the crop weed free early, for two, three, four or five weeks, was better than leaving the crop weedy for five weeks and weeding from then onwards.

Trials on inter-row cultivation, with a row crop cultivator, a spider gang tool or a brush hoe, did not provide adequate weed control in sweet corn, nor did flexi-tine harrowing. Cultivations must be shallow to avoid root damage. Flame weeding has given short-term weed control but could not maintain control of germinating weeds through the season (*see* Weed Control In Sweet Corn above). Directed burners and the use of leaf protectors reduces crop injury. It is also possible to undersow sweet corn, after the first hoeing operation, with white or red clover or trefoil, which can help to suppress weeds and establish a subsequent ley.

Weed Management in Umbelliferas

Umbellifers, like carrot and parsnip, can be difficult and slow to establish and are generally uncompetitive in the initial period of establishment. Early weed control and the use of stale seedbeds will, therefore, give good results.

CARROTS

In drilled carrots, left weedy through to harvest, yield losses can be up to 100 per cent. Even low numbers of weeds can have a serious effect and weed biomass (or weight) gives a better measure of potential crop losses than weed density. Carrots have little ability to recover after removal of weed competition and so keeping the crop weed free, for a period of up to forty-two days after sowing, is necessary to prevent yield loss.

Hand weeding may be cost effective in organic carrots but costs for between 100 and 300 person hours per hectare (40 to 120 person hours per acre) for hand weeding, have been recorded on holdings. Stale seedbed cultivations, flame weeding or brush weeding and other inter-row cultivations can reduce the need for hand weeding and are more economic in the long term. On land with a low weed population, pre-emergence flaming followed by a single weeding, at three to five weeks after crop emergence, have been sufficient to prevent crop losses due to weeds. For post-emergence weed control, it was demonstrated in one study that a steerage hoe was more successful than a flame or a brush-weeder, under the trial conditions in which the weed flora was relatively tolerant of flaming and a hard soil-cap made the brush-weeder ineffective.

Covers (applied for early emergence or carrot fly control) also increase weed growth up to four-fold and additional weeding may be needed under protected crops. If covers are removed, they should be kept off for a short a time as possible to stop pests entering the crop or, preferably, only be removed when forecasting indicates that pest numbers are low.

Fig 91 Clean carrots after a brush weeding treatment.

CELERY
Celery is usually transplanted into beds and is slow to get away. The early stages of growth do not, therefore, provide much shading of weeds, which, coupled with the high moisture regime required following transplanting, favour weed grow. Weed management will depend on establishing a rotation that reduces weed burden and the effective use of stale seedbeds.

PARSNIP
These are slow to emerge and slow to develop, therefore, early weed control is very important. As with carrots, the use of stale seedbeds and pre-emergence flaming, followed later by mechanical weeding, can aid weed control. Late sowing can increase speed of emergence and help when using stale seedbeds

Weed Management in Grassland and Leys

It is difficult to make an economic evaluation of the benefits of controlling weeds in grassland for a variety of reasons. For example, the so called grass weeds differ little from the desirable grasses and they have some feed value. Any extra production achieved through weed control measures only has a benefit if it is then utilized more intensively. Economic evaluation is also complicated because grass is not widely traded, as such, and its value is the contribution it makes to livestock production. For a livestock enterprise, the important thing is to balance grass production with demand. In addition to providing grazing for stock, grassland is also the source of hay and, together with other forage crops, a constituent of silage.

Newly Sown Grass–Clover
Careful preparation of the land and rapid establishment of the crop are the first, and most essential, requirement for weed free, young leys. It is important that perennial weeds, such as docks and thistle, are destroyed before the seed is sown because these weeds are difficult to control in established grass. The land should be ploughed as deeply as practicable, to bury any perennial weeds to the maximum possible depth. In the spring, the land should be worked to a fine, moist seedbed. If sowing can be delayed until July or August, there will be more time to clean the land at a time when weed seed germination is relatively low. However, sowing should not be so late that the clover does not establish before the winter. After spring establishment, topping will help to control annual broad-leaved weeds but must be carried out before the weeds seed. Weeds in the

established sward are controlled by chain harrowing in spring or regular topping through the growing season. In some circumstances, leys may be established by undersowing in a cereal crop, but timing is critical to avoid out-competing the cereal, while ensuring good establishment of the leys.

Established Grass

It is sometimes difficult to get agreement on what is a weed in grassland. A species that is a weed at certain times of year, or in one type of grassland, may be of value at a different time of year or in a different situation. The species composition of established grassland can be influenced markedly by different systems of management. Reseeding of an established pasture to improve the vigour, by increasing the species composition or introducing clover without destroying the existing sward, may done by broadcasting the seeds alone or in slurry, after first cutting the grass for silage.

In established swards, chain harrowing in spring, levels and aerates the soil surface. Regular grazing and cutting will aid weed control. The grazing regime will depend on the management and condition of the pasture, and on the livestock. Sub-soiling with the mole plough and other sub-surface implements, aerates the soil and improves drainage without breaking up or inverting the sward. The surface can also be aerated with spiking machines or using implements with blades that cut slits into the turf. To prevent poaching, stock should be kept off swards on heavy land in very wet conditions, heavy machinery should only be used when conditions are good and gateways and feeding areas should have hard standing or materials that give the soil additional support.

Upland Grass

This is characterized by poorer grass species and is at risk from invasion by heather and bracken. Rushes may be a problem in high rainfall or poor drainage conditions. The grassland may be improved but this may not be economically justified, or be undesirable from a conservation point of view. Weed management will normally be by livestock, often sheep.

Weed Management in Legumes

Beans such as field beans and runner beans can be an important component of both arable and vegetable rotations. Weed management will obviously depend on the specific situation in which crops are being grown and the ultimate market. Weeds are a particular problem in crops grown for processing. At crop harvest, easily uprooted weeds can contaminate the crop with soil and stones carried on their roots. Other weeds become entangled in the harvesting machinery. Berries, flower heads and seed capsules can contaminate the final product, unless removed by flotation or by hand.

Weed Management in Field Beans

Autumn sown field beans are generally resilient to weed interference. In some field studies, the maximum yield loss was only 33 per cent; these also showed that the time of weed emergence, relative to the crop, is also important, as the earlier the weeds emerged, the greater the level of yield reduction.

The cultivations needed for seedbed preparation depend on the weeds present and the time of sowing, but will normally involve similar preparations to other arable crops. Disking may be required if there is a high weed population present. Later sowings demand a finer seedbed and may involve using spring tines. There may also be time to include a stale seedbed approach. Broadcasting the seed, or using narrow row spacing, gives greater weed suppression but reduces the weed control options. Wide row spacing facilitates mechanical weed control. Weed control may be initially with spring tine harrows, then inter-row hoeing with the number of passes depending on weed density and crop and weed growth stages. Blind harrowing and inter-row cultivations can keep the weeds under control at early crop stages. Once the early weeds are controlled, the field beans usually smother the later emerging weeds out.

Weed Management in Runner Beans

In 'pinched' runner beans left weedy through to harvest, yield losses can range from 0 to 98 per cent. This crop is, generally, a poor competitor. Even low numbers of weeds can have a serious effect, although weed biomass gives a better measure of potential crop losses than weed density. The land should be free of couch, creeping thistle, docks and other perennial weeds, and standard seedbed preparation methods, like preparing a stale seedbed, should be used to reduce weed numbers before the crop is transplanted.

Weeding is complicated by the growth habit of the crop and the fact that supports are needed in the field. After crop establishment, hand hoeing is often the most efficient weeding method, especially close to the plants. Mulching with black plastic, or a similar material, is possible on a field-scale, with runner beans planted through holes in the sheeting; this can suppress weeds around the crop. However, the system is expensive due to the costs of material and labour. As this is a tall crop, it is also possible to undersow with a low growing legume, which can suppress weed growth, but some trailing will be necessary to avoid competition.

Mulching can be combined with the use of crop covers for earlier cropping. Hand hoeing in May takes forty-five to ninety person hours per hectare (eighteen to thirty-five person hours per acre). In runner beans in Poland, a single weeding, between four and seven weeks after sowing, was sufficient to prevent a significant yield loss (where, without any weed control, yield loss could reach 90 per cent).

Weed Management in French or Dwarf Beans

Green dwarf beans needed to be kept weed free for up to six weeks from sowing to prevent loss of yield. In drilled dwarf beans left weedy through to harvest, yield losses have reached 90 per cent. As with runner beans, French beans are a poor competitor, especially in early crop stages and a high weed biomass can affect yield considerably. In dwarf bean cultivars, weed fresh weights are less under cultivars with a branching habit (indeterminate varieties) that provide a dense leaf canopy, as compared to erect (determinate) varieties that have limited branching. On the bonus side, French beans can be transplanted when the plants are quite large and can be quick to crop, especially if picked for green beans.

In studies, closer row spacings of 15–46cm (6–8in) reduced weed biomass, as compared with wider spacing of 91cm (36in), and produced higher yields. Weed suppression was not enough to prevent a significant yield loss, compared with the weed free crop. Where the crop was weeded initially, further weed growth was reduced substantially by the narrower row spacings.

Mechanical weeding is possible in some situations to control weeds between rows, but not within rows, and crop injury can be a problem, as pods are generally exposed. French beans cannot be harrowed from three days prior to emergence until the first true leaf stage, because the loop of the emerging shoot is easily broken. In trials, a single pass with various weeders in snap beans did not provide adequate weed control. Machines included a row crop cultivator, a spider gang tool and a brush hoe. Flexi-tine harrowing did not control weeds either, but using combinations of the implements provided better weed control (but not where rainfall delayed weeding operations). Climatic conditions following soil disturbance, during mechanical inter-row cultivations, determined weed abundance, which is greater in wet, as opposed to dry, conditions.

Weed Management in Broad Beans

Left weedy through to harvest, yield losses in broad beans can be high, ranging between 0 to 80 per cent. The crop can suffer low weed numbers without significant loss, but the composition of the weed flora is important in determining the total weed biomass and hence crop yield loss.

Competition studies between the crop and weeds indicate that a single weeding, at three weeks after crop emergence, should be sufficient to prevent yield loss. In practice, keeping the crop weed free for the first month should suffice to allow the crop to establish and out-compete the weeds. Chain harrowing can be carried out pre-emergence, and later the crop can be ridged with care. Wide row spacings allow inter-row cultivations and hand weeding. It is also important to ensure free air movement around plants to avoid fungal diseases. Weed control might also be important to reduce the numbers of alternate hosts of the black bean aphid (*Aphis fabae*), of which fat hen is one of many summer hosts.

Weed Management in Peas

The effect of weeds on the yield of peas depends on relative times of crop and weed emergence, and differences in seasonal rainfall. Peas that emerge first generally suffer less competition than those that emerge later. In vining peas, increasing crop density reduces the total fresh weight of the weed flora but at higher crop densities, within crop competition increases, and this can negatively affect yield, so that a balance has to be struck between yield and weed suppression. Even at high crop densities, weeds can still reduce yield through competition, while at low crop densities, the weeds reduce tillering of the peas. Weeds also reduce vining throughput during crop harvest.

Fig 92 Weeds can hinder harvesting in peas.

Cultivar choice and seeding rate are important aspects of weed control in organic pea production. The semi-leafless nature of many modern cultivars allows greater light penetration through the crop to the weeds. In tests of the competitive ability of pea cultivars, it was shown that vigorous types, with a high biomass accumulation, had a higher competitive ability. Larger seeded cultivars (such as Zelda and Ambassador) tend to be more competitive than the smaller *petits pois* types and semi-leafless peas.

Seeding rate should be as high as economically possible, 120 plants/m^2 (12 plants/ft) being seen as the minimum. Chain harrows can be used pre-emergence and even post-emergence of the crop, if compensated for by higher seed rates. In vining peas, a single, relatively late inter-row hoeing, when peas had four to nine nodes and weeds were 5–10cm (2–4in) tall, generally controlled weeds better than an earlier treatment. A sequence of an early followed by a later cultivation did not given any better weed control. Additional treatment with a torsion weeder gave only slightly improved weed control.

Weed Management in Oil and Fibre Crops

Organic oil and fibre crops are likely to grow in importance in the future, not only for direct consumption but due the need for oils as feedstock in a range of industries (e.g. cosmetics) and due to the need for biofuels in one form or another. Weed management in these crops is likely to depend, at least initially, on experience in other organic arable crops and, to a limited degree, on experience in managing weeds in the conventionally grown crops.

Oil Seed Rape

In organic crops, broad-leaved weeds are considered to be less of a problem than wild oats and black grass. In a survey of conventional winter oilseed rape in central–southern England in 1985, cleavers was the most frequent weed, being found in 57 per cent of fields, whilst common poppy, prickly sowthistle and scentless mayweed were also common (21, 18 and 14 per cent of fields, respectively). Volunteer barley can have a severe effect in the autumn on the growth of winter oilseed rape and, in Canada, it has been shown that increasing the seed rate, improves crop yield when volunteer barley is a problem but there is no advantage when the crop is weed free.

Linseed/ Flax

This crop is not competitive and cannot compete against common chickweed, charlock, wild oats and other quick growing annuals. Fields should be free of perennial weeds. Experiments to determine the competitive effects of weeds on spring sown linseed found wild oat and knotgrass to be highly competitive. Chickweed and fat hen were less damaging. Chickweed formed a low mat that did not affect the crop growth, while the fat hen did not become as vigorous as it does in many crops. In winter linseed, chickweed and cleavers could be left in the crop until March or April without seriously jeopardising yields. Flax cultivars differ in their susceptibility to the allelochemicals produced by wild oats and some other weeds.

Hemp

There is time, before drilling in late April, to prepare a stale seedbed. Once the crop is established, the foliage is dense and suppresses weeds, minimizing further weed control needs. Hemp has been shown to be a useful break crop in a rotation and could be used to 'clean' weed infested fields, as part of an arable rotation. Weed free crop material is essential for processing.

Weed Management in Potatoes

Weeds reduce crop yield by an average of around 36 per cent but losses can be anything from 14 to 80 per cent. Weeds are not always a problem in potato but control may be considered necessary to safeguard crop quality and yield. Perennial broad-leaved weeds, including creeping thistle, field bindweed, the docks, and the perennial grass weeds, common couch and black bent, are particular problems in potatoes. Among the annual weeds, taller species, such as fat hen, are the most problematic. In the UK, a single weeding, between two and eight weeks after crop planting, has been shown to be sufficient to prevent a significant yield loss.

In potato cultivars, greater competitive ability is associated with early emergence, rapid early growth and the ability to develop a dense leaf canopy. In this respect, chitting the tubers aids early establishment of potato crops. Cultivars may differ both in growth rate and habit, which may affect the time to canopy closure. Some cultivars are known to produce taller and leafier foliage than others. However, mechanical weed control can damage lush foliage, especially later in the season. Increasing planting density may improve weed suppression but it also increases establishment costs and may encourage blight due to the greater density of foliage.

The normal cultural practice is to ridge shortly after planting and let the ridges settle. Weed control is then applied ten days after planting, using chain harrows, ridgers or purpose built weeders. Harrowing down the ridges, between planting and emergence, hastens crop emergence as well as controlling weeds. The number of passes depends on weed density. Where a second harrowing is needed, this may be carried out at, or shortly before, crop emergence. Thermal weed control can also be used to control seedling weeds prior to crop emergence.

Inter-row cultivations between the ridges and re-ridging are carried out, as needed, post crop emergence. Cultivations are best done when weeds are small and unlikely to re-establish. Rolling cultivators, used by some growers, have tines that weed between the rows and rolling star-shaped tines to cultivate the sides of the ridges. The ridges are then rebuilt by ridging bodies that follow the tines. In addition to controlling weeds, earthing up the ridges covers the developing tubers and reduces the incidence of tuber blight. However, post-planting cultivations should be kept to the minimum necessary to achieve weed control, as they also accelerate water loss and can cause direct crop injury. Crop damage can result in an average yield loss of 7 per cent but losses due to weeds can be considerably higher.

There has been some interest in the use of living mulches for weed control. Cover crops such as vetch, oats, barley and red clover have been evaluated for their potential ability to suppress early season weeds in potatoes, where the cover-crops were inter-seeded at ridging or three, four or five weeks after crop planting. Weed control with red clover was consistently poor. The cereals and vetches gave good weed suppression but reduced crop yield. The application of green manure material, from mustard and oilseed rape, can result in weed suppression due to the release of allelochemicals as the mulch decomposes. In the USA, an autumn sown rapeseed green manure crop incorporated in spring prior to potato planting, reduced weed density by 73 and 85 per cent in different years, and reduced weed biomass by 50 and 96 per cent in those years. Mulches of paper, plastic and other materials have given good control of weeds but are economic only in high value early potato crops. Where early potatoes are grown under floating covers, weed control is difficult until the covers are removed.

Chapter 7

Solving Weed Management Problems

In this book we have shown that weeds are important components, and weed management an essential process, within organic farm systems, and that both feature prominently in the day to day running of any farm. This is because they are likely to have direct and usually, but not always, negative impacts on desired farm outcomes if left unmanaged – for instance, reduced yield, reduced financial returns, increased management costs, increased personal stress. However, organic farms are complex biological, social and economic systems, and due to this complexity, weed 'problems' usually have a diverse range of causes and effects. The experience of farmers, growers, advisors and researchers has also shown that, because of this complexity, these problems are unlikely to be solved by a single, simple, technological fix.

In fact, weed management underlies the rationale for many farm operations and many have, at least in part, a weed-management component. On-farm experience has shown that organic weed management, under real farm conditions, requires forward planning combined with a very practical and hands-on approach. Whilst any plan needs to take a whole rotation perspective to managing weeds, with planned crop sequences and weeding operations, it also needs to be flexible enough to accommodate changing situations and conditions as they occur in the field, so that it is possible to take advantage of any opportunities to manage weeds as they arise; for example, dry spells that allow machinery to be used on the field or an unexpected weed flush that can be used to control a group of weeds. An in-built flexibility is also required because weed management is, above all, an on-going process and, therefore, does not have an end point as such. Weeds and weed management form a backdrop to the farm system and it would be difficult to imagine a 'natural' farm system in which they were not present.

Solutions to, or strategies for, managing weeds are, therefore, likely to be multi-faceted and entail an understanding of a farm, farm history and a full appreciation of the farmer's goals. This will involve knowledge of many factors at field level (such as field history, crop, weather), at farm level (such as size, enterprises, location) and at regional level (such as markets, policy, environment). This implies, and experience confirms, that weed management is as much about knowledge development and learning, as about physical field practices. In this chapter we discuss the ways in which farmers and growers can learn to analyse and effect changes in their weed management strategies, to better bring about the diverse goals that they have for their farming practices. We also discuss the ways in which farm advisors and organic researchers can aid this learning process. Finally, we look at the learning approaches and tools that can be used to effectively join together all these actors, so as to allow learning for weed management in organic farming.

CREATING KNOWLEDGE FOR WEED MANAGEMENT

Each organic farm system is in many ways unique. It is the outcome of its location, local physical conditions, past farm history, market outlets and the goals and aspirations of the farmer or grower. Weed management is one aspect of the farm system and its effectiveness and economic viability will depend on all the factors that go to make up the farm situation. An effective organic weed management strategy is likely to need to integrate a range of different management approaches across the whole farm system, and any such strategy is likely to work best over a period of time. Farmers and farm workers are often the best placed people to understand, evaluate and develop the necessarily detailed knowledge to formulate a specific weed management strategy, as it forms the backdrop to their working lives and is of more immediate concern to them directly. As a consequence, farmers and growers (and here we include others, such as farm workers and farm advisors) should be constantly seeking new knowledge on weeds and weed management to adapt to their farm situation and to incorporate into their weed management strategies. This involves engaging in a lot of work that has become associated with agricultural white collar 'professionals', such as weed scientists, agronomists or advisors, and can include taking on many developmental and advisory roles as well as some research tasks. By emphasizing the importance of seeking information and knowledge, such an approach seeks to create opportunities for all those involved to learn together about weed management, produce information that is of direct benefit to farmers and growers in their practical day-to-day concerns about weeds and weed management, and create a beneficial cycle of knowledge development that helps to improve the ability of organic farm systems to be resilient in changing circumstances.

This is nothing new! It has been demonstrated, and is well known, that all farmers and growers do a lot of 'practical experimentation' with weed management on their farms. This takes the form of detailed observations on aspects of weeds or weeding, simple comparisons of different treatments (for example, trying different varieties or sowing rates), as well as modifications to farm equipment (adapting or designing new kit). They do this to develop practical weed-management techniques, adapted to specific situations, which work for them on their farms. In addition to this, most farmers and growers look for opportunities to acquire new knowledge, principally from other farmers. Any programmes aimed at improving weed management strategies should aim to facilitate, support and stimulate this on-farm and practical learning approach. Below, we discuss some of the approaches and methods that might be usefully employed to do this.

Fig 93 Recording and making observations on weeds is a valuable way of learning about them.

However, it should also be recognized that, in parallel to this informal experiential learning system, there exists more or less formal organic research and extension (advisory) networks in most countries. Paid for from a mix of private and public funding, they devote at least some resources to research and advice on weeds, and might even include organic weed management, (in some cases!). The current role of this formal network is perceived as (researchers) generating technologies to resolve farmer problems and then (advisors) transferring these technologies to farmers, with some feedback between the various actors to fine tune the research that is done. Although the output of this system can supply useful knowledge, such as formally describing details of weed biology or specific weed management operations under somewhat controlled conditions, it still needs to be re-developed for use at the farm level because, as previously described, weed management in a specific organic farm system will depend on a large number of contingent and historical factors. Many of these will be difficult for advisors, and especially researchers, to grasp in detail. It is, therefore, difficult to supply strictly relevant research solutions in all situations and, in these circumstances, it is often not clear where formal 'scientific' research can help in specific cases (how many times have you heard researchers or advisors say 'Well that depends...'!). In fact, many weed-management questions are best answered by farmers, advisors and researchers working closely together as colleagues, and learning together, rather than relying on the technology transfer approach described above.

Indeed, it is clear that no farmers or advisors or researchers think that they have all the answers to questions about organic weed management practices. A perhaps surprising outcome of recent weed research projects is that knowledge is as important as the actual weeding operations, when it comes to weed control. 'Knowledge is power' to understand and improve weed management situations on a farm and, more importantly, to effect change, if necessary. Effective knowledge development and learning for weed management, which is a continual and on-going process, should place farmers at the centre of an informal knowledge exchange network to which individuals can contribute and share knowledge. Researchers and advisors are best used as part of such a network and to support on-farm practical research (through providing information, helping stimulate learning, monitoring, and assisting farmers to assess the effects of their management interventions). Knowledge is not simply information, but is the result of perception, experience and reasoning, or learning, which is a crucial element for weed management in organic farming systems and we outline below many of the ways in which farmers can become actively engaged together, with advisors and researchers, in learning about weed management.

LEARNING FOR WEED MANAGEMENT

Learning Needs

Experience has shown that effective weed management on organic farms is about learning to improve complex situations and that learning approaches should encourage farmers, advisors, researchers, and any other relevant actors, to jointly develop knowledge about weeds and weed management and, crucially, to share it. The important point is to define and engage all actors in the process in a way that will benefit all. Understanding the different needs of the various key players is also important when pursuing a common cause. Below, we have attempted to describe some of the key needs that the principle actors might have in any programme that is facilitating the development of weed management strategies on organic farms.

Farmers and Growers
This group seeks specific (often detailed) information on 'problem' weeds, and are especially interested in how other farmers are managing their weeds. They make observations on weeds and they undertake a great deal of informal experimentation with weed management techniques and methods in their own fields, and combine all this knowledge in order to develop practical weed management methods, applicable in their farm situation. This knowledge is often not 'formally' evaluated in a scientific sense. It is used to modify weed management practices, which are subject to a continuous developmental process within the farm system.

Researchers
Researchers formulate detailed research questions (or hypotheses) about weeds and weed management methods. However, for the reasons described above, many 'weed problems' are often difficult to define 'scientifically', and may be situation specific, so that many scientific trials will not be grounded in addressing immediate practical concerns. Researchers generally undertake detailed replicated trials (normally of limited scope) that are statistically evaluated and can often be used to produce over arching 'recommendations' of general applicability. They need to do this to achieve recognition within their own professional circles. However, once again it should be recognized that, because of the complexity involved, many experimental results can at best only address a few of the likely key factors and thus the results may be open to many different interpretations. As a consequence, farmers and advisors would normally seek to add practical detail to much of the knowledge generated!

Advisors
These are normally perceived as acting as go-betweens and, potentially, have a more flexible and less defined role. When paid from the public purse, they are traditionally seen as taking information from scientific research and adapting it to provide messages that farmers can use on their farms. However, advisors may also be allied to businesses or other interest organizations (for example, organic certification bodies) or, increasingly, paid for by farmers themselves and, in each case, the message would be expected to be tailored to the funder to a greater or lesser extent. In many cases, researcher information might be by-passed or judged irrelevant. In the better cases, advisors would also be expected to transmit messages from farmers to researchers, although mechanisms for doing this are usually much less well defined.

Farming Press
An often overlooked group interested in knowledge development is the farming press, which is more or less well developed in the UK. The farming press has an interest in selling copy and, therefore, engages to a greater or lesser extent in generating newsworthy stories that interest farmers and advisors, some of which can be innovative. In a general sense, journalists will be driven by the need to generate interesting stories that will maintain circulation but they can also be constrained by column inches and the need to provide concise 'sound bites'. It can be extremely productive to engage the professional press in knowledge development and it can certainly ensure dissemination of knowledge to a wide audience, but it should be recognized that sometimes elements of the message or story can get lost.

Other Actors
A large number of other potentially interested parties might have a stake in specific weed management programmes. For instance, conservation bodies increasingly buy and manage farms to promote a particular end; for example, to maintain populations of rare birds or butterflies. Even where conservation bodies do not directly engage in farming, they can have an interest, sometimes statutory, in preserving or maintaining specific sites of national importance or interest. In all these cases, weed management practices can directly impact on the aims of conservation programmes. Increasingly, consumers and other public interest groups are also engaging in farming through environmental and/or health issues. In these cases, the ways weeds are managed can have a bearing on the way in which people perceive a situation and how they are likely to react when buying produce. It is, therefore, increasingly necessary to engage other

'stakeholders' in weed management programmes and to understand their 'needs' from a wider perspective.

Learning Approaches

In the case of organic weed management, we have argued for engagement of all the principal actors (farmers, researchers, advisors) as a collaborative team in a learning approach. Some specific team-building tasks that have been found useful in a learning approach to weed management are given below.

Collection and Collation of Existing Knowledge

This is necessary in order to provide an underpinning to the development of resilient weed management strategies. It is important that information is collected from all relevant sources, which can include, among others, research papers, notes from farm walks or advisor technical bulletins. Such information should be held in a way that it can be easily shared and easily evaluated, thus forming a solid base for the generation of relevant knowledge when needed. Just as important, the information should be in a form that can be used to generate and develop new information and knowledge.

Monitoring Weed Management Practices

All actors are interested in monitoring weed management strategies and evaluating their effects. Although finding common assessment criteria can be a challenge, such information can be rewarding and help all actors to understand how management practices can be adapted and changed to improve farm situations. In organic farming systems, it is particularly important to try and monitor weed management strategies across rotations, as information from specific parts of rotations may be misleading if taken in isolation.

Monitoring Farmer Experiments

All farmers engage in on-farm experimentation. Engaging all actors in such practices, although challenging, can also be rewarding. Systematically altering factors and analysing effects is a powerful way of rapidly building up knowledge about systems and, when done in a wide range of situations, can help to more accurately define system responses and genuine research questions that need to be answered.

Learning Tools

A range of learning tools can be used to engage farmers, advisors and researchers in developing learning approaches to weed management. We list and discuss some that have proved effective, although we would stress that they are not the only ones. There is also a great deal of cross-over between the various methods outlined below. The important point is that we would urge all farmers, advisors and researchers to engage in such processes when they have the opportunity, as a stimulus to their professional interests and as an aid to acquiring knowledge about weeds. You never know, you might learn something!

Field Walks

These are probably the most popular and, potentially, the most useful way of engaging all actors. Field walks should be centred around a specific weed management theme. It can be constructive to visit different farms that are tackling similar problems at the same time or in a series of visits. The aim of farm walks should be to create a dialogue between farmers, advisors, researchers and other interested parties. Farm walks can also be used to resolve issues around weed management by talking through situations and understanding how they have come about – a co-learning

Fig 94 Discussing weed management in potatoes on a field walk.

approach. Although it is sometimes difficult to stick to specific topics, this is not necessarily a weakness, as it allows many different facets of weed management to be explored. It can be a challenge to record the outcome of farm walks but this can be necessary to draw other stakeholders into the discussion.

Farmers and Focus Groups
Such groups are another popular form of stimulating learning. Groups can be formed around particular 'problems' (e.g. weed management groups), geographical locations and business improvement. More normally, they might include all of these elements and more. Structured discussion around weeds, which includes advisors and researchers, can greatly facilitate learning and help farmers adapt weed management strategies. They can also lead to other forms of learning, such as farm walks and experimentation. An important function of these groups can be to lobby for effective change in other areas of concern to farmers, such as at the political or market level.

Workshops, Open Days, Seminars and Meetings
These are all organized by a range of organizations as an aid to information transfer or exchange. They range from a more active engagement of participants in defining and discussing problems (typically workshops) to a more passive absorption of information (typically seminars and meetings). The best workshops can involve a mixed range of participants discussing specific problems, often involving reflection on specific situations and working to understand them through (for example) cause–effect–problem–solution analyses, mapping and/or other forms of diagramming. They naturally lead to all participants taking discussions back to, and as a result modifying, their working practices. Seminars, meetings and open days are often arranged to transfer knowledge about specific issues or research projects and can be useful for acquiring a large amount of information quickly, although this may need to be reflected on, or revisited, at a later time to be effectively absorbed.

Case Studies
Elaborating case studies on weed management practices is an effective tool for aiding learning. Case studies can concentrate on specific weeding practices or widen out to describe whole farm approaches. It is important to revisit case studies from time to time to understand why weeding practices have changed and to create a deeper understanding of weeding on organic farms. Case studies can also be effectively combined with farm walks, if group members are willing to engage in this activity.

Experimentation
Workshops and meetings can often lead to questions that are not easily resolved or to which answers are not known. In such cases, more or less simple experiments can be used as an aid to learning about such problems. Trials can range from simpler farmer led field trials, which try to resolve practical issues or choose between a limited number of choices, to more complex researcher led trials that may be needed in situations where there is very limited knowledge and a large number of factors that need to be teased out. In any case, involving advisors and researchers in solving farmers own research questions can be a powerful and quick way of building up a body of relevant knowledge that will feed into other areas, such as farm walks and group meetings.

Literature Reviews, Books, Leaflets
The written word is still a powerful way of conveying information and is still a primary resource for information on weeds and weed management. It is often presented as leaflets, popular press articles and scientific papers. The source of such information is from scientific reviews of weed research relevant to organic weed management and synthesis of information held by farmers, advisors and researchers. Many sources of this information exist and should be sought out in answering specific questions (*see* Appendix for information sources).

Fig 95 Judging the effectiveness of robot guided hoe in a trial in wheat sown in rows.

Internet

The internet is increasingly becoming an important database or repository for much information held on weeds and organic weed management. It has the advantage that it is cheap to store such information and it is, potentially, easy to access. The possible disadvantage is that not all farmers are comfortable with the various technologies involved. Increasingly, websites are becoming available that store knowledge in an available form and provide pointers to other sources of knowledge, such as: organicweeds.org.uk. Information exists on individual weeds, weed management, including crop strategies, case studies of weeding strategies on organic farms and even commercial sites selling weeding equipment.

Exchange Visits

Visits between interest groups can be an effective way of stimulating a rapid learning atmosphere, as people are exposed to different practices and attitudes to weed management. This can be particularly effective if the visits are arranged to different countries.

USING INFORMATION AND KNOWLEDGE

Finally, it is true to say that there is no lack of knowledge about weeds and organic weed management of relevance to organic farming. In fact, many farmers report that they are often overwhelmed by the amount of different information available. In such cases it becomes important to have a system for storing and recalling information on weed management. In the end, it will be a matter of individual choice and personal preference in how this is done but here we provide a few pointers:

Keep farm records and notes. It is important to keep good farm records on weeds and weeding operations so that any changes or unexpected occurrences between crops or seasons can be picked up and potentially explained. It is also important to be able to judge whether a weed is becoming more prevalent or decreasing. It should be remembered that this should be evaluated from a rotational point of view. Records should include an evaluation of costs associated with weeding, so that judgements can be made about the costs as compared to the benefits. Such notes will be invaluable when it comes to judging the effectiveness of weeding operations over a long time period and, in this respect, notes on weather conditions can also help. Modern recording methods can make such a task much less onerous and the use of digital cameras can help to record weeds and weed presence, and when laid side by side, help to describe situations.

Keep up to date with current developments. The farming press regularly carries stories on weed research, the outcome of weed management research and even case studies of individual farms. Increasingly, the internet also reports such material online. It is worth keeping cuttings of relevant news items on weeds and weed management and building up a file of such material. It is much easier to leaf through a file than hundreds of back issues, which normally end up on the recycling heap anyway! Material obtained at open days and workshops can also be added to a file, as can leaflets, notes from farm walks and other printed material. However, do not worry if, during busy periods, you miss things; information has a habit of coming around many times. It is productive to occasionally try and match some of this information with farm records and notes and discard any irrelevant material.

Keep on learning and experimenting. Weed management is, above all, a process and weed problems will change from crop to crop, season to season, and rotation to rotation. Within this pattern it will be necessary to plan (using farm notes and current developments as a guide) to manage weeds but it will also be necessary to take opportunities as they arise. An

Fig 96 The answer to your weed problem might be over the fence.

open, flexible approach to weed management, incorporating a diverse range of practices, carried out over the long term, is more likely to lead to an effective weed-management programme, which maintains weeds at acceptable levels. Such an attitude to weed management is also more likely to lead to an appreciation of the place of weeds in an organic farm system and lead to a more satisfying programme that blends in with other aspects of the farm system and farm goals.

Appendix: Useful Information For Weed Management

PERIODICALS

A range of periodicals are aimed at farmers, growers and advisors. They are a quick way to keep up with the latest thinking on weed management, although they contain a lot of other information as well. In addition, many certification bodies, research institutes, levy bodies and agricultural colleges publish newsletters or occasional bulletins on a wide range of farming issues, including weed management. Many of the mainstream press periodicals are mainly aimed at conventional farmers, although they all carry stories and news items on organic farming and growing as well. It is also worth keeping an ear out for farming programmes on local radio, as well as national radio and television.

Clover – Quarterly magazine of the Organic Trust in Eire (tel: +353-1-853 0271).

Commercial Grower – Weekly press and news dedicated to (mainly conventional) UK growers (tel: 01322 612112).

Farmers Weekly – Weekly press and news (tel: 0845 0777744).

Farmers Guardian – Weekly press and news (tel: 01858 43883).

Horticulture Week (including *Grower*) – Weekly magazine on horticulture (tel: 020 8606 7500 or hortweek.com).

Organic Farming – A technical magazine for producers produced quarterly by the Soil Association (certifiers and lobby group) (tel: 0117 929 0661).

Organic Matters – Bi-monthly newsletter on organic farming in Ireland. Irish Organic Farmers and Growers Association.

Organic Research Centre Elf Farm Bulletin – Research updates and regular technical, management and financial advice for producers (tel: 01488 658298).

Organic Studies Centre Technical Bulletin – A quarterly update from the Organic Studies Centre, Duchy College, Cornwall. (tel: 01209 722148).

Organic Today – Quarterly journal for producers from Organic Farmers and Growers Ltd. (tel: 0845 330 5122).

Soil Association Certification News Bi-monthly Newsletter – For all Soil Association certified producers. Regular updates on production standards issues and market contacts service (tel: 0117 914 2406).

Star and Furrow – Twice yearly journal for the Biodynamic Agricultural Association (tel: 01453 759501).

The Organic Way – A quarterly newsletter on organic gardening (tel: 024 7630 8200).

Vegetable Farmer – A monthly magazine for growers (tel: 01622 695656).

BOOKS, LEAFLETS AND TECHNICAL GUIDES ON WEEDS AND WEED MANAGEMENT

ADAS Colour Atlas of Weed Seedlings. Williams, J.B and Morrison, J.R. (Mosby-Wolfe, 1987)

Arable Plants a Field Guide. Wilson, P. and King, M. (English Nature/Wildguides, 2003)

Growing Green – Organic Techniques for a Sustainable Future. Hall, J and Tolhust, I. (The Vegan Organic Network, 2006, 328pp.)

Field Guide to the Wild Flowers of Britain. Press, J.R., Sutton, D.A. and Tebbs, B.R. (Readers Digest Association Ltd, 2002)

Flora of the British Isles. Clapham, A.R., Tutin, T.G. and Moore, D.M. (Cambridge University Press, 1990)

Grasses: A Guide to Their Structure, Identification, Uses and Distribution in the British Isles. Hubbard, C.E and Hubbard, J.C.E (ed.) (Penguin Press Science,1988)

Mechanical Weeding in Organic Production Systems: How, Why, When. OPICO Ltd (Advanta and Elm Farm Research Centre, 2000)

New Flora of the British Isles, 2nd edn. Stace, C. (Cambridge University Press, 1997)

Organic Cereals and Pulses. Weed Control in Organic Cereals and Pulses. Davies, D.H.K. and Welsh, J.P. (available at orgprints.org/8162)

Organic Farm Management Handbook, 7th edn. Lampkin, N., Measures, M. and Padel, S. (2007)

Organic Farming. Lampkin, N. (Old Pond Publishing, 2002)

Organic Farming and Growing. Blake, F. (Crowood Press, 1994)

Organic Vegetable Production – A Complete Guide. Davies, G. and Lennartsson, M. (ed.) (Crowood Press, 2006)

Practical Weed Control in Arable and Outdoor Vegetable Cultivation Without Chemicals. Schans, D.A. *et al.* (Wageningen University and Research Center Publications, 2006)

Profitable Organic Farming. Newton, J. (Blackwell Science, 1995)

Steel in the Field a Farmer's Guide to Weed Management Tools. Bowman, G. (ed.) (Sustainable Agriculture Network Handbook Series, no 2) – A book incorporating a large amount of farmer experience with mechanical weeding tools; also containing case studies, although some are more suited to the US rather than the UK (available online for free at sare.org).

Technical Guides from the Soil Association – A range of crop specific guides that contain information on producing various crop organically including information on weed control. (contact Soil Association Producer Services (tel: 0117 314 5000 or soilassociation.org). They include:

Improving Biodiversity on Organic Farms
Organic Carrot Production
Organic Grassland Management and Forage Conservation
Organic Livestock Management on Nature Conservation Sites
Organic Onion Production
Organic Potato Production
Pig Ignorant?
Rotations for Organic Horticultural Field Crops
Soil Management on Organic Farms
Weed and Scrub Control on Organic Grassland

The Photographic Guide to Identity Garden and Field Weeds. Phillips, R. (Elm Tree Books)

The Wildflower Key. Rose, F. (Penguin Books, 1981)

The Wild Flowers of Britain and Northern Europe, 5th edn. Fitter, A., Fitter, R. and Blamey, M. (HarperCollins, 1996)

Weed 'Em and Reap – A two-part DVD series showcasing tools and reduced-tillage strategies for non-chemical weed management in which growers and researchers describe field-tested methods for managing weeds without the use of chemicals. Available from the Oregon State University (c/o Weed 'Em and Reap, OSU–Dept of Horticulture, 4017 Ag and Life Sciences Bldg., Corvallis OR 97331-7304, USA).

Weed Guide. (Bayer Crop Science, 1995) – An expert guide to arable weeds and weed control.

Weed Identification Guide. HDC (Horticultural Development Council, 2006)

Weed Management Technical Leaflets are available from HDRA (all available free on-line at organicweeds.org). Including:

Annual Weed Management in Organic Systems
Bracken Management in Organic Systems
Charlock Management in Organic Systems

Couch Management in Organic Systems
Creeping Buttercup Management in Organic Systems
Creeping Thistle Management in Organic Systems
Dock Management in Organic Systems
Fat-Hen Management in Organic Systems
Field Horsetail Management in Organic Systems
Ragwort Management in Organic Systems
Rush Management in Organic Systems
Allelopathy – A Practical Weed Management Tool?
Minimum Tillage – Is it a Viable Option For Organic Systems?
Weed Management in Organic Systems – Fallowing
Weed Management in Organic Alliums
Weed Management in Organic Brassica's
Weed Management in Organic Carrots
Weed Management in Organic Cereals
Weed Management in Organic Fruit
Weed Management in Organic Lettuce
Weed Management in Organic Potatoes
Weeds and Weed Management on Arable Land. An Ecological Approach. Hakansson, S. (CABI) – A comprehensive overview of the ecological approach to managing weeds in arable crops.

INTERNET INFORMATION ON WEED MANAGEMENT

The internet now contains a large amount of information on weeds and organic weed management. There are also a large number of sites that enable identification of weeds and give advice on their management. Some information should be verified with different sources before being acted upon. Below we have listed some of the more established websites that may aid in gaining access to some of this information.

A range of agricultural colleges offer training and courses in subjects pertinent to organic weed management and access to this information can also be obtained on the internet.

There are a large number of weeding machinery manufacturers, agents or sellers and it is beyond the scope of this book to list them. They can be easily found with a search on the internet. There are also a large number on auctions and sales, where it is possible to buy good second-hand weeding kit. Sales are normally listed in the farming press or associated websites.

Danish Centre for Organic Food and Farming (DARCOF) – A range of material on weed management including access to research through the organic e-prints database (darcof.dk and orgprints.org).
Elm Farm – Some published research on weeds (efrc.com).
HDRA Organic Weed-Management Website – All aspects of organic weed-management, including an extensive database on individual weeds, weeding methods and organic farmer case studies (organicweeds.org.uk).
National Sustainable Agriculture Information Service (ATTRA) – A large amount of on-line information for organic farmers and growers, aimed at US farmers (attra.ncat.org).
Organic Soil Fertility Project – Information on cover and fertility building crops (organicsoilfertility.co.uk).
Scottish College of Agriculture – Information and factsheets on organic weed-management (sac.ac.uk).
USDA Alternative Farming Systems Information Centre Agriculture – Access to technology for alternative farming systems in the USA (nal.usda.gov).
Weedspotter – An on-line identification guide to weeds for UK from Bayer Crop Science, aimed at conventional farmers and therefore with lots of advice about herbicides not relevant to organic farmers and growers but useful none the less for the pictures (bayercropscience.co.uk).

USEFUL ADDRESSES

Weed events are a good way to learn about weeds and weed management. A number of organizations are engaged in organizing and holding regular national and regional events for organic producers (not all aimed at weed management!). The main ones are listed below. Some are membership organizations and many publish a regular list of events. Advice, training and courses on organic farming and growing are also available from many of the organizations listed. Many also have web pages with information on weeding.

Abacus Organic Associates – The UK's leading group of independent organic consultants.
c/o Rowan House, 9 Pinfold Close, South Luffenham, Rutland, LE15 8NE
(tel: 01780 721019; e: enquiry@abacusorganic.co.uk; internet: abacusorganic.co.uk)

ADAS (Agricultural Development and Advisory Service)
Woodthorne, Wergs Road, Wolverhampton, WV6 8TQ.
(tel: 0845 766 0085; internet: adas.co.uk)

Association of Independent Crop Consultants
Agriculture House, Station Road, LISS, Hants GU33 7AR
(tel: 01730 895354)

Biodynamic Agricultural Association (BDAA) – For biodynamic farmers and growers.
c/o The Secretary, Painswick Inn, Stroud, GL5 1QG
(tel/fax: 01453 759501; e: office@biodynamic.org.uk; internet: biodynamic.org.uk)

Central Science Laboratory
Sand Hutton, York YO41 1LZ
(tel: 01904 462000; e: science@csl.gov.uk)

Centre for Organic Seed Information – Organic seed database.
(internet: cosi.org.uk)

Farming and Wildlife Advisory Group (FWAG) – Offers many training events and farm walks of interest to organic growers and farmers.
(internet: fwag.org.uk; see website for regional offices)

HDRA – Research Department of Garden Organic. Organizes events mainly for horticultural producers and gardeners
(e: research@hdra.org.uk. internet: gardenorganic.org.uk)

Holm Lacy College c/o Holme Lacy, Hereford, HR2 6LL
(tel: 01432 870316; fax: 01432 870566; e: holmelacy@pershore.ac.uk; internet: projectcarrot.org)

Home Grown Cereals Authority (HGCA)
Caledonian House, 223 Pentonville Road, London N1 9HY
(tel: 020 7520 3926)

Horticultural Development Council (HDC
Bradbourne House, East Malling, Kent ME19 6DZ.
(tel: 01732 848383; e: hdc@hdc.org.uk; internet: hdc.org.uk)

Horticultural Research International Association (HRI-A)
c/o Wellesbourne (Headquarters), Warwick CF35 9EF
(tel: 01789 470382; e: hri.association@hri.ac.uk; internet: hri.ac.uk)

Institute of Grassland and Environmental Research (IGER)
c/o Plas Gogerddan, Aberystwyth, SY23 3EB
(tel: 01970 823026; e: iger.reception@bbsrc.ac.uk; internet: iger.bbsrc.ac.uk)

Irish Organic Farmers and Growers Association (IOFGA) – Have a number of monitoring farms where farm walks and demonstrations are held on a regular basis.
c/o Main Street, Newtownforbes, Co. Longford, Republic of Ireland
(tel: +353 (0) 43 42495; e: iofga@eircom.net: internet: iofga.com)

LANTRA – The Sector Skills Council for the Environmental and Land-based Sector.
Lantra House, Stoneleigh Park, Nr Coventry, Warwickshire CV8 2LG
(tel: 024 7669 6996; e: connect@lantra.co.uk; internet: lantra.co.uk)

Lazy Dog Tools Ltd
Hill Top Farm, Spaunton Appleton-Le-Moors, Yorkshire YO62 6TR
(tel: 01751 417351; e: enquired@lazydogtool.co.uk; internet: lazydogtools.co.uk)

Machinery Rings Association
c/o RAMSAK Ltd, weald Granary, Seven Mile Lane,
Mereworth, Maidstone, ME18 5P2
(tel: 01622 815356; e: info@machineryrings.org.uk; internal: machineryrings.org.uk)

Nafferton Ecological Farming Group
c/o Nafferton Farm, Stocksfield, Northumberland, NE43 7XD
(tel: 01661 830222; internet: ncl.ac.uk; e: tcoa@ncl.ac.uk)

National Association of Agricultural Contractors
8 High Street, Maldon, Essex CM9 5PJ
(tel: 01621 841675)

National Institute for Agricultural Botany (NIAB)
Huntingdon Road, Cambridge CB3 OLE
(tel: 01223 276381; e: info@niab.com)

North East Organic Programme
c/o PO Box 321, Newcastle upon Tyne NE3 2YP
(tel: 0845 121 7645; e: neop@northeastorganic.org; internet: northeastorganic.org)

Northwest Organic Centre – c/o Rural Business Centre, Myerscough College, Myerscough Hall, Bilsborrow, Preston PR3 0RY
(tel: 01995 642206; e: enquiries@nworganiccentre.org; internet: nworganiccentre.org)

Organic Arable Marketing Group
Elm Farm Research Centre, Hamstead Marshall, Newbury RG20 0HR
(tel: 01488 657600)

Organic Centre (Ireland)
c/o Rossinver, Co. Leitrim, Ireland
(tel: +353 71-98-54338; fax: +353 71-98-54343; e: organiccentre@eircom.net; internet: theorganiccentre.ie)

Organic Centre Wales – Organizes a demonstration farm network, events and training courses regularly at venues throughout Wales.
c/o University of Wales, Aberystwyth, Ceredigion, SY23 3AL
(tel: 01970 622248; e: organic@aber.ac.uk; internet: organic.aber.ac.uk)

Organic Conversion Information Service (OCIS)
(tel: England 0117 922 7707, Scotland 01224 711072, Wales 01970 622100, Northern Ireland (crops and horticulture) 028 9070 1115 (livestock) 028 9442 6752)

Organic South West – Promotes organic farming in Cornwall and Devon. It provides producer information and training, and organizes sector groups.
Kyle Coberparc, Stoke Climsland, Callington, Cornwall, PL17 8PH.
(tel: 01579 371147; e: osw@soilassociation.org)

Organic Studies Centre – Hosts many farm walks events and training courses.
c/o Duchy College, Rosewarne, Camborne Cornwall TR14 0AB
(tel: 01209 722155; internet: organicstudiescornwall.co.uk)

Scottish Agricultural College (SAC) Organic Farming
c/o Ferguson Building, Craibstone Estate, Bucksburn, Aberdeen, AB21 9YA
(tel: 01224 711072; e: David.Younie@sac.co.uk; internet: sac.ac.uk)

Soil Association Scotland
c/o 18 Liberton Brae, Tower Mains, Edinburgh, EH16 6AE
(tel: 0131 666 2474; e: contact@sascotland.org)

The Institute of Organic Training and Advice (IOTA) – Provides information, training and support to specialist organic advisers and trainers.
c/o Cow Hall, Newcastle, Craven Arms, Shropshire SY7 8PG
(tel: 01588 6640118; e: iota@organicadvice.org.uk; internet: organicadvice.org.uk)

The Organic Advisory Service and Organic Centre Elm Farm – Holds regular meetings and farm walks in England for converting and

established organic farmers, covering dairy, beef and sheep, arable and horticulture.
c/o Hamstead Marshall, Newbury, Berkshire, RG20 0HR
(tel. 01488 658 279; e: gillian.w@efrc.com; internet: efrc.com)
The Organic Demonstration Farm Network of Elm Farm Organic Research Centre – Hosts events (seminars, farm walks and training days) in England.
(e: education@efrc.com; internet: efrc.com)

The Soil Association – Produces a list of events throughout England, Wales and Scotland.
c/o Producer Services, South Plaza, Marlborough Street, Bristol BS1 3NX
(tel: 0117 314 5000; e: info@soilassociation.org; internet: soilassociation.org)

World Wide Opportunities on Organic Farms WWOOF
PO Box 2675, Lewes, East Sussex, BN7 1RB
(tel: 01273 476 286; e: hello@wwoof.org.uk; internet: wwoof.org)

Yorkshire Organic Centre – For events in Yorkshire and the Humber region.
c/o Skipton Auction Market, Gargrave Road, Skipton, North Yorkshire BD23 1UD
(tel: 01756 796222; e:info@yorkshireorganiccentre.org; internet: yorkshire-organiccentre.org).

Index

Allelopathy 11, 29, 91, 94, 199, 210, 216
Annual meadow grass 83, 148, 198, 214
Annual weeds
  broad-leaved 16, 117, 205
  grasses 17, 148
  management 16, 83, 114–117

Barren brome 43, 150
Bent 49, 177, 178, 182
Biennial weeds 17, 155
  management 154
Bindweed 11, 44, 117, 173
Biological weed control 68, 90, 111
  broad spectrum 93
  classical 91
  conservation 92, 157, 167, 168, 172
  innundative 92
Black grass 43, 48, 151, 214, 220
Bracken 113, 185, 239
Break crop 28, 198, 244
Broad-leaved (weeds) 16, 117, 160
Broadspectrum control 66, 71
Buttercup 160, 162, 163

Charlock 49, 119
Cleavers 11, 120, 213
Chickweed 45, 48, 49, 122, 198, 209, 213, 220
Composting 54, 58
Corn Spurrey 36, 124
Costs 33, 62, 65, 68, 95, 98, 207
  cereals 105
  grass 106
  recording/calculating 98–101
  vegetables 103
Couch 18, 44, 49, 67, 83, 179–182, 220
Cover crop 28, 51, 88, 94, 198
Creeping soft grass 182
Crop
  choice 25,199, 210, 218, 220
  covers 226, 237
  damage 212, 226
  establishment 37, 75, 200
  management 18, 33, 208–245
  sequence 24–26
  spacing 34, 36, 201, 218, 221, 230
  variety 33, 199, 212, 218, 221
  weeding costs 103–107
Critical (optimum) periods 65, 116, 209, 220
Cutting (see topping) 39, 72, 205

Direct (physical) weed control 61, 68, 70, 96, 201
Docks 17, 20, 36, 39, 44, 58, 67, 70, 107, 164–168, 197, 200, 203
Dormancy 15, 41, 115

Energy use 109
Environmental costs 108–112

Fallowing 32, 160, 167, 174, 180
Fat Hen 44, 48, 83, 126, 209
Flame weeders 82, 215
Fumitory 123

Gallant Soldiers 128
Grass weeds 83, 148, 177
Grazing 51–53, 204
Groundsel 43, 49, 91, 129

Hand weeding 68, 96, 98, 105, 207, 232, 237
Hoes 69, 74
Horsetail 36, 113, 186
Hygiene 37, 55, 114, 217

Infra-red burners 84
Inter-cropping 30, 199, 216, 224
Inter-row (weeding) 66, 74, 202, 223
Intra-row (weeding) 66, 79

Knotgrass 44, 83, 131
Knowledge gathering (record keeping) 21, 55, 98, 201, 247, 252, 256

Labour costs 96
Learning 21, 226, 247, 250, 257
Leys 26, 203, 222, 238
Livestock 51, 58, 93

Machinery 56, 68, 70–81, 207, 211, 215, 219, 224, 231
  brush weeders 77, 98, 101
  costs 97, 98–100
  cutters 72
  finger weeders 81
  harrows 72, 201, 211, 217, 222
  mounted hoes 74
  mowers 72
  ploughs 40
  rotovators 45, 76–77
  tine weeders 45, 72, 81, 101, 202
  torsion weeders 81
  steerage hoes 74, 78, 231
Management Strategies 18, 23, 61, 64, 115, 225
  alliums 228
  arable 105, 197, 208, 210–226, 239, 243
  beet 208, 209, 228
  brassicas 228
  grass 106, 203, 238
  legumes 235, 239–243
  maize 214, 236
  vegetables 88, 89, 103, 205, 226–238
  oil and fibre 243
  potatoes 234, 244
  salads 233
  squashes 234
  umbelliferas 236
  wheat 218
Manures 53, 58, 205
Mayweed 43, 134, 136, 209, 213, 220
Mechanical weed control (see machinery)
Mulch 68, 87, 94, 216, 225, 245
  living 28, 88, 245
  sheeted 89, 98, 235
  particle 88

Nettle 141, 168, 205
Nightshade 44, 118
Non-cropped areas 58

Onion couch 44, 183
Organic standards 19, 57

Pasture 38, 106, 203
Patch weeding 66, 67, 70, 203
Penny-Cress 127
Perennial weeds 32, 42, 69, 159, 177, 185
  creeping 17, 159
  management 160
  stationary 17, 159
Persicaria 132, 133
Pineapple weed 137
Ploughing 40
Poppy 44, 138, 140
Principles
  organic 19
  of weed management 20, 23, 62, 64, 111, 197

Ragwort 20, 44, 70, 74, 93, 107, 155–157
Red Shank 133
Rotation 24, 45, 103, 111, 197, 226
Rushes 36, 188–191, 239

Seed
  cleaning 57
  rate 35, 201, 219, 221, 223
  quality 34, 56, 115, 200
Shepherd's Purse 43, 49, 83, 140, 198
Slurry 53, 58, 205

Soil
  condition 12, 36, 40, 71, 201, 209
  nutrients 13, 36, 55, 222
Sow-thistle 44, 142, 143, 175
Spear thistle 17, 20, 157
Speedwell 44, 49, 144, 145, 146, 213
Stale (false) seedbed 45–47, 115
Statutory regulation 19, 156, 159, 164, 170, 200

Thermal weed control 68, 82–86, 215
  weed sensitivity 83
Thresholds 62, 102, 201
Thistles 18, 20, 37, 44, 58, 70, 74, 158, 170–173, 203
Tillage 39, 42–44,
  in dark 47
  post harvest 48–50
  primary 40, 200, 209, 222
  reduced (or zero) 41, 50
  ridge 50
  secondary 44, 200, 209, 226
  tertiary (see direct weed control) 61, 70
Timing (of weeding) 26, 45, 64, 116, 205, 211
Topping 27, 72, 106, 204
Transplanting 37, 229, 230

Undersowing 31, 88, 167

Variety (choice) 33,
Volunteer weeds 18, 42, 113, 191–195
  beet 195
  cereals 192
  oilseed rape 193
  potato 194

Weather 71, 246
Weed
  benefits 12, 60, 107–108
  ecology 14, 18, 103
  negative effects 10, 62, 197
  seed bank 16, 41, 114
  traits 15
  types 16, 43, 61, 113, 159
Wild Oat 27, 43, 153, 198, 201, 203, 214
Wild radish 147